Evolution of Broadcast Content Distribution

Roland Beutler

Evolution of Broadcast Content Distribution

 Springer

Roland Beutler
Südwestrundfunk (SWR)
Stuttgart, Germany

ISBN 978-3-319-83411-5   ISBN 978-3-319-45973-8   (eBook)
DOI 10.1007/978-3-319-45973-8

Printed on acid-free paper

This Springer imprint is published by Springer Nature
The registered company is Springer International Publishing AG
The registered company address is Gewerbestrasse 11, 6330 Cham, Switzerland

Dedicated to Kai and Chris

Acknowledgments

New ideas and insights thrive and prosper on vibrant and inspiring exchange of views. The FDS project group at the EBU was a great environment to look at broadcasting from a different perspective which paved the way to new understanding.

I am very grateful to all colleagues for their engagement, enthusiasm and passion. My special thanks go to Darko for all his support, backing and help to find the right direction over the last years.

Contents

List of Figures

List of Tables

List of Acronyms

APT	Asia-Pacific Telecommunity
ASMG	Arab Spectrum Management Group
ATU	African Telecommunications Union
AVMSD	Audio-Visual Media Service Directive
BNetzA	Bundesnetzagentur
CDN	Content Delivery Network
CEPT	Conférence Européenne des Administrations des Postes et des Télécommunication
CITEL	Comisión Interamericana de Telecomunicaciones
DSL	Digital Subscriber Line
DTT	Digital Terrestrial Television
DVB	Digital Video Broadcast
DVB-T	Digital Video Broadcast–Terrestrial
DVB-T2	Digital Video Broadcast–Terrestrial 2
EBU	European Broadcasting Union
EC	European Commission
ECC	Electronic Communications Committee
ECO	European Communications Office
EFIS	ECO Frequency Information System
eMBMS	enhanced Multimedia Broadcast Multicast Service
EPG	Electronic Programme Guide
ETSI	European Telecommunications Standards Institute
EU	European Union
FCC	Federal Communications Commission
FOBTV	Future of Broadcast Television
FTA	Free-To-Air
FTTH	Fibre-To-The-Home
FTV	Free-To-View
GEO	Geostationary Orbit
HAPS	High-Altitude-Platform-Systems
HbbTV	Hybrid Broadcast Broadband Television

HDTV	High Definition Television
HEO	Highly Elliptical Orbit, High Earth Orbit
HPHT	High-Power-High-Tower
IEEE	Institute of Electrical and Electronics Engineers
IMT	International Mobile Telecommunications
IoT	Internet of Things
IP	Internet Protocol
ISD	Inter-Site Distance
ISP	Internet Service Provider
IRT	Institut für Rundfunktechnik
ITU	International Telecommunication Union
IXP	Internet Exchange Point
LEO	Low Earth Orbit
LPLT	Low-Power-Low-Tower
LTE	Long Term Evolution
M2M	Machine-to-Machine
MEO	Medium Earth Orbit
MFN	Multi Frequency Network
MNO	Mobile Network Operator
NFV	Network Functions Virtualization
Ofcom	Office of Communications
OTT	Over-The-Top
PoP	Point of Presence
PSB	Public Service Broadcasting
PSM	Public Service Media
RCC	Regional Commonwealth in the Field of Communication
RR	Radio Regulations
RSC	Radio Spectrum Committee
RSD	Radio Spectrum Decision
RSPG	Radio Spectrum Policy Group
SDN	Software-Defined Networking
SDTV	Standard Definition Television
SFN	Single Frequency Network
SVT	Swedisch Television
TFA	Table of Frequency Allocations
UHDTV	Ultra High Definition Television
UPU	Universal Postal Union
VOD	Video-on-Demand
WRC	World Radiocommunication Conference
3GPP	Third Generation Partnership Project

Chapter 1
Introduction

Broadcasting has come a long way since the first radio programmes were transmitted about a century ago. It developed into a solid pillar of societies in terms of informing, entertaining, and educating the population. However, the world has completely changed with the onset of the digital revolution some decades ago. Obviously, this did not leave broadcasting untouched.

The broadcasting sector can roughly be subdivided into three main areas. Those are content generation, distribution, and consumption. Contemporary technologies have further split these three areas into more branches. New players have entered the scene and therefore traditional broadcasting companies are getting increasingly confronted with severe competition. Their former dominating role of basically the only providers of audio-visual content for mass audiences is steadily crumbling.

This is particularly true for public service broadcasters which sometimes have to fulfill a mission impossible. On one side they are meant to serve every citizen. This includes providing content attractive to a mass audience, on the other side they shall not forget about the different niches. The content shall be of high quality with regard to ethic and moral standards. And, at the same time audience ratings have to be high.

If public service broadcasters do not reach all social milieus, critics say they do not fulfill their remit. If they engage in too popular programme formats, it is argued that this is not their business. Furthermore, public service broadcasters are subsidized by public money and usually are not allowed to make profit. Therefore, they are not market participants like commercial broadcasters which are fully profit oriented.

Public service broadcasters are also under pressure from policy makers. There are trends in some countries to cut their independence from politics as well as attempts to undermine the position in general terms. There are politicians who question the

© Springer International Publishing Switzerland 2017
R. Beutler, *Evolution of Broadcast Content Distribution*,
DOI 10.1007/978-3-319-45973-8_1

necessity of public service broadcasting. They claim that the USA as an example would show that market forces are enough to create an independent and prosper broadcasting sector. Even though this is certainly an exciting area for debate, this issue will not be further pursued here.

In a globalized media market which is more and more dominated by the development and the potential of the Internet the situation is getting tighter for public service broadcasters in terms of maintaining their market share. They are forced to offer more services, better quality of services, and last but not least new types of services. This is a threat and an opportunity at the same time.

Integration of social media has without any doubt lifted the attractiveness of public service broadcasters to a new level. However, social media come along with new habits and expectations of users, they are using new types of devices and they want to have access to any kind of broadcast content wherever they are, at all times, under all conceivable receiving conditions and certainly at costs which are affordable.

For broadcasters this is directly linked to the question of content distribution. More clearly, it is about getting all content to all devices at any time under the conditions broadcasters have to adhere to. The latter includes confining and controlling the costs.

In the past distribution of broadcast content was straightforward. In order to distribute linear programmes broadcasting networks could do the job. With the increasing demand for interactivity and the wish of users for more autonomy in their media consumption broadcasters had to look for new means of delivering their content.

Meanwhile the situation has become very confusing. Therefore, broadcasters have to put more effort into finding a solution for their content distribution through which they can achieve their objectives. Content distribution thus is of very high strategic value and broadcasters need to develop their own strategies towards it.

This book gives an overview about the development of broadcast content distribution. Moreover, it tries to sketch those elements and factors which have direct bearing on strategic considerations and decisions relating to content distribution. At the time of writing this text (2016) there are several important activities underway, in particular with respect to technological developments. Also, the regulatory framework of the telecommunication sector is being scrutinized currently in Europe. All this will definitely have an impact on future broadcast content distribution.

As a start, Chap. 2 gives a high-level description of the broadcast ecosystem from content production to its consumption on the user side. After that, distribution options which are currently available for broadcasters are sketched in Chap. 3. Chapter 4 delves into the regulatory framework under which distribution of broadcast content takes place, i.e. who are the main organizations and how they are interrelated. This leads to an overview about current technological development trends. This is addressed in Chap. 6.

The focus of this book lies on public service broadcasting. Therefore, in order to properly analyze and assess future decisions regarding distribution it is necessary to clarify the particular interest of public service broadcasters. Most of their requirements are directly linked to the regulatory and economic conditions under which they have to operate. This is summarized in Chap. 7.

Distribution is of strategic importance for broadcasting companies. This is more than ever true in a quickly evolving environment driven by the innovations of broadband communication. In Chap. 8 elements are identified and approaches are put forward which may support broadcasters in their effort to define their future distribution strategy. Finally, Chap. 9 summarizes the main points and suggests some areas where broadcasters should engage or increase their effort.

Chapter 2
The Broadcasting Ecosystem

For most people broadcasting is a synonym for receiving and enjoying many different types of radio and television programmes. Over more than a century it has been the primary source of audio-visual content. Even though not visible for the public, the broadcasting industry encompasses a great variety of different branches which contribute to the production and distribution of radio and TV programmes. The broadcasting value chain begins with those who produce content, i.e. in the first place radio and TV programmes, and ends with those who deliver and offer this content to users. Figure 2.1 sketches the different areas along the value chain.

The first element in the broadcasting ecosystem is the creation of audio-visual content. This includes radio shows, movies, documentaries, news for radio and TV, etc. These are the classical content offers of broadcasters. However, with the advent of the Internet new types of content have emerged such as websites, blogs, social media appearance, and so forth. In the meantime these new forms developed into an indispensable pillar of the content offer of modern broadcasting companies. For the discussion presented here, the terms "content" or "broadcast content" is meant to address the entire offer of a broadcasting company.

Broadcast content is offered to users in terms of a broadcast service. This means a set of different elements such as movies, shows, news, sports, etc. is bundled and offered as a package. The major part of broadcast services are radio or television programmes. However, a broadcast service may also contain non-traditional components as for example on-demand content or social media.

When talking about broadcast services it is important to make a distinction between linear and nonlinear content. Linear broadcast content or service refers to the traditional way of offering radio or TV services. Listeners and viewers tune in to the scheduled sequence of content and consume what they are offered over a certain period of time. The sequences of radio and TV programmes are set up by broadcasters and cannot be changed by a listener or a viewer. The only user interaction with regard to linear services is to change from one service to another

© Springer International Publishing Switzerland 2017
R. Beutler, *Evolution of Broadcast Content Distribution*,
DOI 10.1007/978-3-319-45973-8_2

Fig. 2.1 The broadcasting
value chain

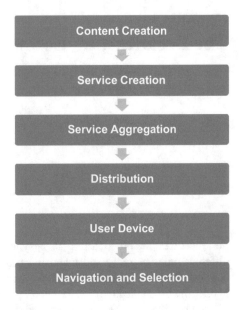

or to switch off if the programme is not attractive. Linear broadcast services are not restricted to traditional broadcasting distribution technology. For example, a live stream on the Internet is to be considered as a linear service as well.

On the contrary, nonlinear content or services require some level of user interaction beyond switching on and selecting something. Typically, the user can select individual pieces of content and control, as a minimum, the timing and sequence of the consumption. Particularly popular nonlinear services are time-shifted and catch-up services. They offer the consumption of content after the start of the live transmission, either while it is still on air (time-shifted) or at a convenient time later (catch-up). Typically, catch-up services are available for a certain period of time, for example a week depending on regulatory and economic constraints.

Other forms of nonlinear broadcast services encompass downloading content to local storage for future consumption or on-demand access to audio and video content for immediate consumption. Furthermore, associated offers such as dedicated websites or data services supporting particular programmes fall under the category of nonlinear services, too. The outstanding characteristic of nonlinear services is the autonomy they offer to the user to decide what to consume, where to consume it, when and on which device.

Content creation is one of the primary tasks of broadcasting companies around the world. They produce news, music shows, and programmes for children, shoot movies and documentaries, and create daily soaps. However, they are not the only creators of content. A huge movie industry, such as the famous studios in Hollywood, produces premier content not only for cinemas but also deliberately for making them available through broadcasting companies. With the digitization of audio-visual content production more and more new players enter the market.

Companies such as Netflix or Amazon spend money on content production themselves with the intention to monetize their products on their own behalf.

The next step in the broadcasting value chain is bringing together more than one service in order to be able to offer a bundle of broadcast services. Many broadcasting companies are actually doing exactly this. Indeed, public service broadcasting companies in Europe are usually offering more than one service. The BBC in the UK has several TV channels providing content 24 h a day, seven days a week. So does the Italian RAI or the German public broadcaster ARD.

On the other end of the chain there are the users which employ very different devices to consume audio-visual content. Traditionally, consumption of broadcast content required usage of a broadcasting receiver, both for radio and TV. Still today there are dedicated broadcasting receivers for fixed, portable, and mobile reception. However, broadcast services are offered over broadband networks as well. In this case, the receivers are either computers or personal devices such as smartphones or tablets.

User devices are getting more and more comfortable. This refers in particular to the support regarding navigation and selection of content. Electronic programme guides are widespread. On personal devices access to broadcast content is often governed by dedicated software applications called apps.

The link between the content side and the user side is established by the distribution mechanism. This is the main topic of the analysis presented here. Before the Internet age broadcast services were distributed by means of dedicated broadcasting networks,[1] i.e. terrestrial, cable, or satellite networks. Today also broadband networks, both wired and wireless are used to carry broadcast content.

Broadcasting or broadband networks are operated by a network operator who sells capacity on the network to content providers in order to carry their products to the users. For example, a broadcasting company such as SWR which is one of the German public broadcasters has a contract with a satellite operator which distributes a defined number of TV programmes through satellite transmission across a large area. The network operator does not provide services himself.

Usually, broadcasters distribute their content over several networks. In order to fulfill their public mandate or to support a particular business model, broadcasting companies select a combination of distribution possibilities which suits them best. Some broadcasters may make their content available through any broadcasting or broadband network, while others choose to use only a subset, for example a terrestrial broadcasting network and satellite, but do not engage with cable distribution.

Whenever the discussion will touch upon a choice of different technical ways to distribute content the term "distribution option" will be used. This refers to the

---

[1]Whenever the transmitter network is meant the term "broadcasting network" is used in order to avoid any misunderstandings. In some regions of the world "broadcast network" can also be used when talking about the network. However, in the USA, for example a broadcast network is usually understood as a broadcasting company such as ABC or NBC.

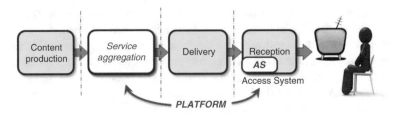

Fig. 2.2 Definition of a distribution platform

physical network or the network technology itself. This is quite often mixed in the public discussion with the term "platform" or "distribution platform" where the word platform is also used to describe the network itself, like talking about "the terrestrial platform" and meaning terrestrial networks in general terms.

For the discussion here the term "platform" will be used to describe something else. According to that, a platform consists of two elements, i.e. a service or a set of services in conjunction with a piece of hardware or software necessary to access the service (see Fig. 2.2). This means the distribution mechanism is not needed to build a platform. Hence, a platform provider does not need to operate a physical network over which content is delivered. There are platforms on traditional broadcasting networks such as Pay-TV platforms on cable or satellite. Well-known Internet platforms are, for example, Netflix or Apple-TV.

Along the broadcasting value chain as shown in Fig. 2.1 there live a huge number of different companies. There are those providing and aggregating content followed by others which create services based on their own content, sometimes complemented by external products. The distribution part sees many different types of networks and infrastructure providers. A plethora of manufacturers of transmitter equipment or network infrastructure are trying to sell their products to the network operators. Last but not least, device manufacturers of various kinds are trying to get their share. With the increase of Internet proliferation the market of audio-visual content has become a center of focus also for companies which were concentrating on broadband services only such as smartphone and tablet manufacturers.

For many players within the broadcasting ecosystem the term "broadcaster" is used. This holds in particular for broadcasting companies whose business is to produce content, create services, and aggregate them. Their services, though, are distributed by other companies in most cases. The BBC [BBC16] is a prominent example for such an arrangement.

However, among broadcasters there are also those doing the same but in addition operating their own terrestrial distribution networks, for example FM or DVB-T/T2 networks. Some of the German public broadcasting companies fall in this category such as SWR [SWR16].

Then, there are companies which solely operate terrestrial broadcasting networks in order to distribute the content of broadcasting companies. Also those companies consider themselves as broadcaster. And finally, sometimes even manufacturers who produce radio or TV receivers see themselves as broadcasters.

Clearly, the interests of all these "broadcasters" are not 100 % congruent. For the sake of clarity, in the following discussion the term broadcaster is only used for broadcasting companies which create content and services irrespective of whether they happen to still operate their own networks or not. Others will be distinguished by calling them explicitly network operators or manufacturers.

Following this logic, there are two main types of broadcasters which are public service broadcasters and commercial broadcasters. Public service broadcasters (PSB) are supported by public money and special regulation (see Sect. 7.1). Most commercial broadcasters finance their services through advertisements in their programmes. However, there are more and more commercial broadcasters which offer their content on pay-per-view or subscription basis.

Chapter 3
Distribution Options

Content and services provided by broadcasting companies have to be transported to the users in order to be consumed on their devices under different conditions. There are several technical options to deliver all kinds of content. They can be divided into two categories which are broadcasting or broadband networks. The former constitute the traditional way of audio-visual content distribution while the latter are usually associated with IP based networks. This chapter introduces the main technical features of the different networks such as structure of networks and their topology. However, technical details will be only given to the extent which is necessary to understand the relevance of the different distribution options with respect to the delivery of broadcast services.

3.1 Broadcasting Networks

There are three basic types of broadcasting networks. Those are terrestrial, cable, and satellite networks. Their primary common characteristic is to distribute a signal from a single distribution point to an in principle unlimited number of concurrent users. There is no connection back from any user device to the source of transmission. In that respect, broadcasting networks constitute a one-way distribution mechanism. In the context of mobile broadband networks this type of communication is quite often also called "downlink-only."

3.1.1 Terrestrial Broadcasting Networks

Historically, terrestrial networks were the first networks which were built to distribute audio and television programmes. This traces back to the end of the 19th

© Springer International Publishing Switzerland 2017
R. Beutler, *Evolution of Broadcast Content Distribution*,
DOI 10.1007/978-3-319-45973-8_3

and the beginning of the twentieth century. Terrestrial networks consist of a set of transmitters which transmit a radio wave using a suitable frequency, bandwidth, and modulation scheme that carries the audio-visual content. At a remote location a user may receive the radio signal by means of a corresponding receiver to which an appropriate antenna is attached. Terrestrial television transmissions started by the end of the 1920s both in the USA and Europe.

Even though the term terrestrial broadcasting network is usually understood as the set of transmitters which distribute the programmes to the user it is clear that the signal has to be transferred to the transmitter sites before it can be broadcast. Depending on the circumstances this can be accomplished by making use of fixed lines including fiber, fixed radio links, or satellite links.

Until the end of the twentieth century terrestrial broadcasting was analogue broadcasting. Then, a new digital transmission technology was provided. In different regions of the world different digital terrestrial television (DTT) broadcasting standards were developed. In Europe DTT broadcasting is based on the ETSI standards for DVB-T [ETS09] and DVB-T2 [ETS15]. Japan and parts of South America rely on ISDB-T [ARI05] while North America uses ATSC [ATS07]. Clearly, there are also standards for different types of digital terrestrial sound broadcasting.

Switch-over from analogue to digital terrestrial broadcasting is a global trend. Most countries in Europe but also the USA, Korea, and Japan accomplished this in the meantime. In Europe broadcasters introduced DVB-T and are now entering the switch-over to the next generation of digital terrestrial television broadcasting by rolling out DVB-T2 networks. This will offer significantly more capacity per multiplex due to the more efficient coding and compression technologies integrated in DVB-T2. Hence, the number of programmes will increase. Furthermore, higher quality can be offered in terms of HDTV programmes instead of only SDTV.

In Europe, Africa, the Middle and the RCC countries digital terrestrial broadcasting is governed by the GE06 Agreement of the ITU [ITU06a] which has been set up by the two Regional Radiocommunication Conferences RRC-04 [ITU04] and RRC-6 [ITU06] in 2004 and 2006, respectively. For both digital sound and television broadcasting a frequency plan has been established together with corresponding rules to change the Plan and bring into operation transmitting stations. This frequency plan uses the frequency ranges 174–230 MHz and 470–862 MHz. Depending on the national situation digital terrestrial television networks offer between 10 and sometime more than 50 different programmes in a given area.

Terrestrial broadcasting networks possess some characteristics which make them very attractive for broadcasters. Their coverage area can be tailored to meet the needs of broadcasters or to adhere to regulatory obligations. Public service broadcasters in particular often have to provide national and regional services at the same time. Broadcasting networks can be rolled out with national or regional coverage.

In the case of digital terrestrial broadcasting systems such as DVB-T/T2 the technique of single frequency networks (SFN) can be employed. The underlying OFDM modulation is the key to this. In an SFN a set of selected transmitters is transmitting the same content on the same frequency. In areas where the signal of a single transmitter would not be strong enough to provide good reception, constructive superposition of signal contributions from different transmitters may overcome the problem.

Operation of an SFN also offers the possibility to easily close gaps in the coverage area by adding a gap filler transmitter in an appropriate location. However, the locations and technical characteristics of the transmitters have to be chosen carefully. Quite often optimization algorithms of various kinds are used to design the broadcasting network. A detailed description of the methods can be found in [Beu04].

Furthermore, using a single frequency across an extended area may be a more efficient use of spectrum than using a different frequency for each transmitter. The latter will lead to a so-called multi-frequency networks (MFN) which used to be the standard way of frequency usage for analogue networks. More details about network planning of digital terrestrial broadcasting systems can be found in [Beu04] and [Beu08].

At the beginning of the broadcasting era, services on terrestrial broadcasting networks were received by means of fixed roof-top antennas. This is still the primary reception mode for broadcast services over terrestrial networks. However, portable and mobile reception can be enabled easily as well. This is a matter of network planning considering denser networks and adapted signal strength levels at the point of reception. Apart from the increased effort and certainly higher costs there is no technical obstacle preventing to implement portable or mobile reception.

Terrestrial broadcasting networks offer several benefits for broadcasters. From a public service broadcasting point of view the most important certainly is that it enables free-to-air (FTA) delivery of broadcast content. Indeed, in Europe it is the only distribution technology which carries FTA programmes in each country. There are countries where satellite also offers free-to-view (FTV) services but these are the minority. A good overview about the availability of FTA services through different distribution options can be found in [EBU14].

Terrestrial broadcasting establishes a one-to-all communication. This means that within a given piece of spectrum a certain number of different broadcast services are offered to an unlimited number of users. The quality of service is controlled by means of proper network planning and is not affected by the size of the audience, i.e. the number of concurrent users. At any given instant of time each user has access to the total capacity of the distribution network. What does not exist in terrestrial broadcasting networks though is a return channel. This means that if some level of interactivity is required, independent means of communication need to be utilized. Hence, broadcasting networks do not allow for the provision of on-demand content nor enable direct feedback to the broadcaster.

Deploying broadcasting networks is not just a technical challenge; it also costs a lot of money. However, achieving given coverage objectives both with regard to area and service quality is independent of the size of the audience. This gives the advantage that the more users are actively receiving broadcast content through terrestrial broadcasting networks the smaller the total network costs per user become. As a consequence, in countries where there is a high market penetration of terrestrial broadcasting as a primary means to receive broadcast services, terrestrial broadcasting is definitely the cheapest way to deliver linear broadcast content. In countries where penetration is low it is just the other way round. Italy is an example for very high usage of terrestrial broadcasting while Germany corresponds to the latter case.

One of the shortcomings of terrestrial broadcasting is certainly that the total amount of available spectrum is pretty limited. This limits the number of programmes which can be offered. Clearly, depending on which broadcasting technology is actually employed spectrum can be exploited more or less efficiently. Also, whether SD or HD programmes are carried and how much bit rate is foreseen for each of them has a significant impact on the total number of programmes carried across one TV channel. In order to be a viable and competitive distribution option digital terrestrial broadcasting requires a sound spectrum basis which can offer enough content to attract users. Therefore, in most countries only the programmes much in demand are distributed across terrestrial broadcasting networks. Niche programmes are usually offered through other means.

Terrestrial broadcasting technologies are adopting only slowly the IP protocol [InP16] to encapsulate the data stream. Rather, they make use of MPEG transport stream technology [TrS16]. Therefore, IP-based receivers cannot be targeted easily by broadcasting networks. This has become one of the hot topics in recent years as broadcasters have realized that it will be increasingly important to be able to reach, for example smartphones and tablets also with their linear services.

As there is an apparently irresistible trend to IP throughout all sectors of communication ATSC has taken the decision to go IP with its new release of ATSC 3.0 [ATS15]. Figure 3.1 summarizes the pros and cons of digital terrestrial broadcasting technologies. Further reading on the value of terrestrial broadcasting can also be found in [EBU14].

In the context of co-primary spectrum allocations of broadcasting spectrum to the mobile service the pros and cons of the terrestrial broadcasting platform have been discussed in length. In the UK a study has been put forward which elaborates on the value of terrestrial broadcasting with an emphasis on economic considerations. The social impact of in particular public service broadcasting on terrestrial networks has been highlighted there as well [Ken14].

An important element of distribution over terrestrial broadcasting networks is that the receiver market is independent from the networks. Buying a DTT receiver together with a corresponding antenna installation is sufficient to receive usually a large number of free-to-air services. This encompasses services of public service broadcasters but also those of many commercial broadcasters. In the case of other distribution options the network operators have influence on the receiving devices or control them entirely.

Fig. 3.1 Pros and cons of terrestrial broadcasting networks

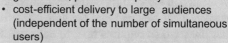

Terrestrial Broadcasting

+
- universal coverage
- any reception mode
- guaranteed, predictable quality
- cost-efficient delivery to large audiences (independent of the number of simultaneous users)
- every user has access to the total capacity of the network

−
- one-way, no return channel
- the offer is limited since capacity is smaller than for other distribution options (no niche channels)
- limited delivery to mobile devices
- no access to IP-only devices
- not cost efficient if penetration is low

3.1.2 Cable Networks

Cable networks which offer TV programmes have been introduced in the second half of the last century. Traditionally, they were broadcasting networks in the sense of providing content from one source to many receivers at the same time. However, the difference between cable and terrestrial networks is that in cable distribution radio signals are not propagating through the air but instead signals are travelling through cables, typically coax cables. Also in the case of cable distribution a certain spectrum range is reserved to carry the TV signals. At the beginning of cable distribution VHF frequencies were used on a regular basis. Nowadays mainly the UHF spectrum is utilized.

In order to distribute a TV channel over a cable network a dedicated frequency is selected. Different TV programmes have to use different frequencies. This has to be done to avoid interference between different programme signals. The approach is similar to terrestrial broadcasting where in a given area either a single analogue programme or a digital multiplex is transmitted on a given frequency. At the point of reception a cable receiver is tuned to the frequency of the selected TV programme which can then demodulate and decode the signal in order to display the television signal on the screen.

In a classical cable network no return channel is provided. This has been overthrown in the meantime by the fact that basically all cable network operators are offering also IP-based access to the Internet thereby naturally establishing a return channel. In the context of this book, however, the term "cable network" will refer to the classical types of networks with no return channel capability.

Distribution over cable networks requires the existence of cable connections from the point where services are generated and aggregated, the so-called head-end to each individual user. At the head-end the cable services are generated by combining programmes from different sources, for example from satellite or terrestrial broadcasting networks. Local content may be added as well as local or regional advertisement.

From the head-end the TV programmes are sent to local feeding points which are located in the vicinity of the residential areas of the subscribers. The connection between local headend and feeding points is nowadays implemented in terms of fiber links which requires fiber to coax conversion at the different feeding points. From there coaxial cables go out to the user homes. This is called a hybrid fiber-coaxial cable television system.

However, in contrast to modern computer networks a cable network does not have a star type topology. Rather, in order to connect several households in a neighborhood a trunk cable is split as often as necessary. Clearly, every time the cable is split the overall signal power is equally distributed over all subsequent connections which calls for repeated amplification. This will eventually decrease the signal-to-noise ratio to the extent that the reception quality will become unacceptable. In this case the cable network operator has to deploy several feeding points which makes the overall network layout more complex. Figure 3.2 visualizes the structure of a classical cable network.

There are some elements of cable networks which discriminate them from terrestrial broadcasting networks. First of all, cable television is subscription based. Users can only get access to services if they subscribe to particular packages of services. Hence, there is no such thing as free-to-air distribution on cable networks. This comes with the fact that even though one can buy a cable receiver in any electronic store it will not come to life without activating the corresponding access control system by inserting, for example, the correct conditional access module. This way the cable network operator controls not only the cable network infrastructure but also the receiver.

According to the definition given in Chap. 2 this actually constitutes a platform, i.e. the cable network provider is offering services and is controlling the receiver at the same time. Since nowadays cable networks not only offer TV services but also Internet access this puts cable network operators in a powerful market position. Indeed, it raises the issue of gate-keeping, in particular for public service broadcasters.

The business relations between broadcasters and cable network operators differ considerably across different countries and regions. There are models in place where broadcasters do not have any direct contractual relations with cable network operators. In these markets cable operators usually generate their services by bundling services of broadcasting companies taken from terrestrial and satellite networks. Such distribution is considered a secondary kind of distribution. In other countries cable operators are paying retransmission fees to broadcasters, or broadcasters are paying carriage fees to cable network operators.

In order to safeguard the availability of particular broadcast content, for example public broadcasting services national administrations have put certain regulation in place (see Chap. 4). Quite often there are so-called must-carry rules which enforce

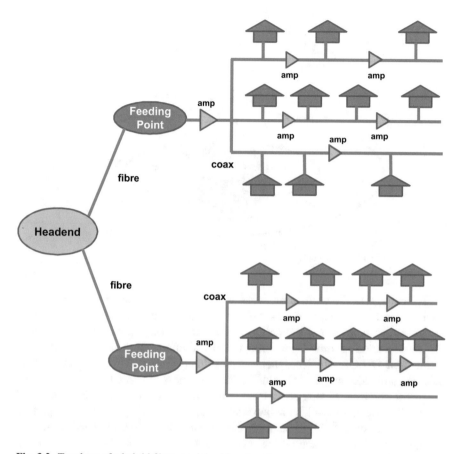

Fig. 3.2 Topology of a hybrid fiber-coaxial cable network

the provision of certain programmes. However, the details of the must-carry rules are usually such that they only guarantee a minimum service offer and not the entire variety of services of a broadcasting company.

3.1.3   Satellite Networks

Since the first satellite has been brought into an orbit around Earth in 1957, i.e. Sputnik 1, satellite networks have developed into one of the basic pillars of global communication. Nowadays thousands of satellites are circling in various orbits. They are used for many different purposes ranging from Earth observation, communication of various kinds to navigation and monitoring weather or carrying out research. Among communication tasks providing broadcast services is also one of the most important usages.

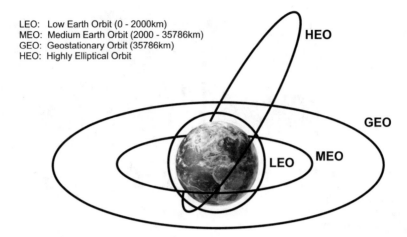

Fig. 3.3 Different types of satellite orbits

Typically, satellite orbits correspond to elliptic curves around the Earth. Some having small eccentricity, others are moving along very eccentric paths. There are many ways to classify satellites. A common way to distinguish them is the classification as Low Earth orbit (LEO), Medium Earth orbit (MEO), High Earth orbit (HEO), and Geosynchronous orbit (GEO) [Sat16]. The latter class also contains geostationary satellites which are most relevant for the delivery of broadcast services. High Earth Orbits quite often are implemented in terms of Highly Elliptical orbits—also abbreviated by HEO—which possess a low perigee in the order of only 1000 km but a high apogee above the geostationary orbit. Very detailed definitions of potential orbits can be found in [Nas16]. Figure 3.3 gives an impression of the difference between the orbit types in qualitative terms.

Depending on the satellite orbit the point which lies on the line between the actual satellite position in space and the center of Earth intersecting the Earth's surface may follow a very complicated path. This means that from the perspective of an observer at some point of the Earth's surface the satellite will describe a complicated curve, too. The satellite may be seen only parts of the day and will change its position in the sky rapidly. As satellites move at very high velocities around the Earth, electromagnetic communication between a fixed station on the ground and a satellite and vice versa has to cope with sometimes severe Doppler shift.

A geostationary orbit is a very special orbit showing favorable characteristics with respect to communication purposes. The intersecting point on the Earth's surface on the line from the satellite to the center of Earth is fixed. Thus, from a location on Earth from which the satellite can be seen a geostationary satellite appears always at the same fixed location in the sky. Hence, there is no need for satellite tracking to adjust the antenna direction correspondingly. The downside of geostationary orbits is their height above ground of approximately 36000 km. This requires powerful transmitters and high gain antennas on both sides, i.e. on-board the satellite and at the Earth station. Also, a signal which is sent over a geostationary satellite link suffers a significant time delay of minimum a quarter of a second which for some applications may be detrimental.

A satellite cruising along its orbit will cover a certain area on the Earth depending on the type of orbit. The coverage area on the ground, which is usually called the footprint of the satellite, will change its location and shape as a function of time. From a given receiving location on the ground satellites may be visible only for a certain period of time. In order to achieve a 24h/7d satellite connection, therefore several satellites have to be employed which are circling the Earth in an orchestrated manner. When the connection to one satellite starts fading due to its changing relative position, another satellite has to take over. The service provision is then handed over from one satellite to the next. This calls for proper orbit planning and synchronization of the satellite network operation.

The situation is less complicated with respect to the geostationary orbit. First of all, the satellite remains at its position all the time relative to the receiving point. This also means that the footprint remains the same both in terms of location and shape. Nevertheless, the coverage area on the ground depends on the performance of the receiving antenna; this is in the first place the diameter of the satellite dish. Figure 3.4[1] gives an indication. Furthermore, a single satellite is sufficient to provide 24h/7d service, i.e. there is no need for any hand-over.

In order to provide a TV service through satellite transmission the signal carrying the TV programmes are fed to a satellite uplink facility. By means of a very large satellite uplink dish, i.e. dish diameters between 9 and 12 m, the signal is transmitted to the satellite in the orbit using a specific frequency range. Large satellite dishes allow more accurate aiming to reach the satellite receiving antennas. Furthermore, the antenna gain increases with increasing dish diameter. This allows reaching sufficiently high signal levels at the satellite.

The uplink signal is received by one of the satellite transponders tuned to the uplink frequency. Then, the signal is shifted to a new frequency range and retransmitted back to Earth. This transmission direction is called downlink. Using different frequency ranges for up- and downlink is necessary to avoid interference between the signals. Many satellites operate in the frequency range 4–8 GHz, the so-called C-Band or between 10–18 GHz, usually addressed as the K_u-Band.

3.2 Broadband Networks

Broadband networks have become an indispensable way for broadcasting companies to deliver linear and nonlinear broadcast services to their users. Quite often the term broadband network is used interchangeably with the term Internet. In order to avoid misunderstandings a distinction could be made by saying that

---

[1] Re-drawn on the basis of [Foo16].

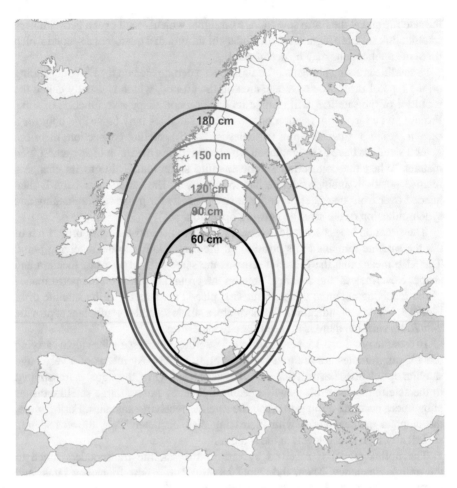

Fig. 3.4 Different footprints of a geostationary satellite targeting at Germany as a function of the satellite dish diameter

the Internet is a conglomerate of different networks of varying architectures, topologies, technologies, sizes and complexity being linked together [ExC16]. The underlying components of the Internet are broadband networks, i.e. different particular technical implementations of network infrastructure which are capable to offer high data rates.

Broadband networks come in two distinct variants. Those are networks which offer best-effort and managed services. Managed broadband services are meant to deliver a guaranteed quality of service and can be employed to deliver linear TV services. Best-effort refers to the standard way of accessing services on the Internet where users usually have to accept varying service quality up to complete failure depending on the actual network traffic (see Sect. 3.2.1). Sometimes this is also called distribution over the open Internet.

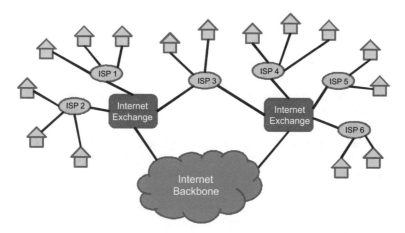

Fig. 3.5 Schematical overview about the Internet topology

The overall structure of the Internet can be subdivided into a backbone or core infrastructure consisting of high capacity data links between geographically distributed Internet nodes and the Internet periphery which enables the access to the backbone networks.

Individuals, enterprises, or organizations of all types and sizes wishing to access the Internet must first of all maintain a connection with their Internet Service Provider (ISP). An ISP runs at least one server that acts as an Internet "Point of Presence" (PoP). This PoP is basically the entry portal through which customers can obtain access to Internet services.

The ISP will access the Internet backbone either directly at an Internet Exchange Point (IXP) or by connecting with another (usually larger) ISP from which they purchase IP transit [Com16]. Figure 3.5 gives a schematic overview.

In general terms, content and services of various kinds are offered by uploading them to a server which is connected to the Internet. Users can access this content through corresponding requests from their Internet enabled devices. The delivery from the server to the user device may pass through several independent networks run by different network operators. These networks are connected by the mentioned exchange points, i.e. IXPs.

Communication over broadband networks is nowadays usually based on the Internet Protocol (IP) [InP16]. Applying IP means that the original data stream is reorganized in terms of creating data packets. These packets are generated according to a particular packet structure that encapsulates the data to be delivered. All packets consist of a header which contains the IP addresses of the source and the destination device and a certain amount of data to be transmitted.

The packets are delivered independently from each other. According to the principles of IP communication [IPS16] packets can take different routes from the source to the destination. Once all packets are received at the destination the original signal can be reconstructed.

There are several basic issues which are of importance here. As different packets can propagate along different routes they may undergo different impacts, also different time delays, and hence some packets may not reach the destination. However, lost packets can be requested again by the receiver until the transferred data is complete at the cost of increasing the overall duration of delivery.

In contrast to this approach, digital broadcasting networks are utilizing a different kind of transport stream mechanism, usually the MPEG transport stream [TrS16]. Even though the technical details may differ from one broadcasting system to another the most important feature of a transport stream approach is that the signal is transmitted as it comes, i.e. the sequence in time is preserved. This remains valid even though there are technologies such as time interleaving which disarranges the temporally correct sequence of bits in a data stream to some extent by changing the order of bits within a given time interval. Such technologies are used to cope with difficult propagation conditions, in particular mobile reception. At the receiver the original bit sequence is reconstructed.

Moreover, since the broadcasting transmission is bound to respect the temporal sequence of the source data it has to be ensured that the transmission is appropriately protected against any impairment along the transmission path. This is usually accomplished by applying efficient error correction schemes in addition.

Apart from the way how data packages are treated, there is one fundamental and very important difference between broadcasting and broadband networks. Broadcasting networks constitute a unidirectional one-to-all communication while broadband networks offer several communication types. The most important is certainly the bidirectional communication between two individual devices or users. This is usually called unicast communication. However, there is also a possibility to implement a broadcast mode similar to a traditional broadcasting network which delivers the same content to an in principle unlimited number of users at the same time. An intermediate type of communication would be multicasting where a limited number of users are served concurrently with the same data stream.

Whatever types of communication are enabled in a given broadband network the total capacity of the network is shared among users. Thus, the higher the number of concurrent users becomes the lower the capacity will be which is available to each of them if all of them are served through unicast connections. This may lead to deterioration of quality, congestions, and consequently to bad user experience. In case there are several users requesting, for example, the same audio-visual content at the same time switching to multicast or even broadcast mode can help to cope with capacity issues.

Most of the communication on broadband networks is unicast communication. From a broadcast content provider point of view it would nevertheless be advantageous to employ broadcast modes on broadband networks as well, in particular with respect to the delivery of linear broadcast services. Naturally, these are meant to be consumed by a large number of concurrent users.

However, the basic problem encountered regarding broadcast or multicast distribution over broadband networks or the open Internet as sketched in Fig. 3.5 is the involvement of many different network operators and ISPs on the way from source to destination. An end-to-end multicast distribution would require cooperation of all parties which is not realistic due to competing business interests and complexity. Consequently, the prevailing form of getting access to services on broadband networks is through unicast connections.

This poses a fundamental issue for broadcasting companies wishing to make use of broadband networks for the distribution of their content, in particular when distribution over the open Internet is envisaged. As long as most traffic on broadband networks is due to unicast connections, distribution costs on broadband networks scale with the number of concurrent user, i.e. the amount of traffic which is generated.

It is certainly true that larger total traffic entails better deals for the price per Mbit between broadcasters and broadband network operators. However, this does not compensate the cost inflation. As current examples clearly indicate the total costs still increase in a way which is critical for broadcasting companies. Swedish Television (SVT) reported in 2015 that their broadband offers adds up to 3 % of the total viewing time of their users while this kind of distribution is responsible for up to 15 % of their total distribution costs [Bjo15]. Similar trends have been confirmed by other European broadcasters, for example the BBC [BBC13].

3.2.1 Over-the-Top and Managed Services

Broadcasting companies are offering most of their programmes online, too. This comprises linear TV and radio services as well as any form of nonlinear content including downloads or time-shifted programmes. There are basically two mechanisms through which content can be made available, i.e. over-the-top (OTT) or as a managed service.

OTT refers to a situation where a content provider creates a website or an app through which his content can be accessed. All broadcasting companies are maintaining such offers. Examples of websites of broadcasters can be found under the URLs www.swr.de [SWR16] or www.bbc.co.uk [BBC16]. Furthermore, there is also a plethora of audio-visual content providers which offer their services through OTT delivery such as Netflix, Hulu, YouTube, just to name a few.

Users having access to the Internet can navigate to these URLs and enjoy the content when they want, where they want and on whichever IP-enabled device they are using. Thus, from a user point of view the only requirement to consume OTT content is to have access to the Internet and—clearly—sufficient bandwidth for a satisfying experience.

The websites of broadcasting companies or other content providers are hosted on dedicated webservers. The webservers can be under the control of broadcasters themselves or operated by an external company. The latter may or may not be an ISP. It can even be a CDN (see Sect. 3.2.5) provider. If broadcasters are utilizing their own servers to host the content, they usually need a contract with an ISP to establish the connection to the Internet, but there are also examples where broadcasters are peering directly at an IXP.

The most important characteristic of OTT delivery of audio-visual content is the fact that the ISPs do not control the content itself nor do they control the quality of the delivery. They are just establishing the means to transmit the IP packets from source to destination. In particular, with respect to the quality of service OTT delivery is a typical best-effort service meaning that users obtain an unspecified variable bit rate and delivery time, depending on the current traffic load.

This is in contrast to a managed service on a broadband network. The idea of a managed service is to overcome the limitations of best-effort offers by ensuring certain parameters relevant for the data transmission. These could be, for example, a certain low level of packet loss or packet delay, security and privacy or a fixed bandwidth. For example, in order to receive a live HDTV signal over a broadband network with satisfying quality the bit rate should not fall below a certain minimum required level.

Managed services are widely available on both fixed and mobile broadband networks (see Sects. 3.2.2 and 3.2.3). Many ISPs offer so-called triple play packages to their customers which contain IP telephony, flat rate Internet access, and a certain number of linear television services. These TV services come with a guaranteed quality of service, so that the customers will always experience the same quality independent of the time of the day or the number of other concurrent Internet users in the neighborhood.

Technically, this is achieved by setting aside part of the overall capacity of the broadband network just for managed services. In essence, this means that the data pipe from ISP to end-user is split in two regimes, one for best-effort services and one for managed services. The decision about how much to allot for both is a central part of the business model of the ISP. It is obvious as every physical data pipe of a broadband network is limited there may be a situation where the availability and quality of service of OTT best-effort services may be affected by the capacity dedicated to managed services. This touches upon the very important issue of net neutrality which has become a crucial regulatory issue for broadcasters (see Chap. 4).

One way of coping with bandwidth bottlenecks in OTT delivery of streaming services is to apply a technique which is called adaptive streaming [AdS16]. It is based on the ability to detect the actually available bandwidth of a user together with the computational power of the receiving device in real time. Depending on the actual situation the quality of a video stream is adapted accordingly. This follows the philosophy that it is more acceptable to users to receive a video stream at a lower quality but still the picture is moving than having to look at frozen pictures while waiting for data to arrive.

The basic idea of adaptive streaming is to provide content stream offers of multiple bit rates. This requires on the server side an encoder which is able to encode a single video source at different bit rates. Then, the user has to employ a video player which can switch between streams encoded differently depending on available resources.

3.2.2 Fixed Broadband Networks

Fixed broadband network refers to a broadband connection to a fixed location. This can be either the home of a user or the premises of an enterprise or an organization. Access to the Internet is established through a router inside the building which acts as the gateway to the ISP. IP-enabled devices inside the building are connecting to the router either by some kind of wired infrastructure or by means of WiFi. In that respect, WiFi can thus be seen as a wireless extension of the fixed broadband network within buildings.

There are three different types of fixed broadband networks. The first exploits existing copper telephone lines offering a broadband service on the basis of a technology called DSL [DSL16]. DSL stands for "Digital Subscriber Line." By using a DSL connection digital data are transmitted over a telephone line by using frequencies not used by voice traffic. Thus, traditional telephony and DSL broadband communication can utilize the same physical line without interfering each other. DSL is probably the most widely used fixed broadband network access in Europe. Depending on the DSL variant, for example VDSL [VDS16] or Vectoring [Vec16], downlink speeds of up 100 MBits/s can be achieved.

The second type of fixed broadband network access becoming more and more popular is offered on the basis of existing television cable networks. For quite some time there is a trend that cable network operators no longer just offer linear television services over their networks. Rather, they started to bundle linear television, telephony, and Internet access into one offer, called Triple Play. Internet access on cable networks is integrated into the cable television infrastructure analogously to DSL which uses the existing telephone network [Cab16]. Downlink speeds can reach up to 400 MBits/s under special conditions.

The most promising way of fixed broadband access to the Internet is accomplished through fiber connections [Fib16]. Communication over fiber networks is not a new technology, indeed the core or backbone networks of the Internet are all connected on the basis of fiber links. However, getting fiber to the homes of users is an expensive undertaking. There have been discussions to do so for quite some time around the planet but so far no large scale roll-out of Fiber-To-The-Home (FTTH) can be detected. Some time ago, Google started investing significantly in fiber networks [Goo16]. After all, there is no doubt that fiber is the technology with the most potential for development, in particular as it seems to be the only fixed broadband network option which may cope with traffic demands of the future.

As mentioned at the end of Sect. 3.2, distribution of broadcast content over fixed broadband networks is very costly for broadcasting companies. However, on the consumer side the cost aspect is usually more favorable. Access to fixed broadband networks is usually provided for a fixed amount of money per month. Flat rate price models are dominating which means there are no data caps, at least for the time being. Hence, users can enjoy online broadcast content without any limits. Limitations may only be encountered due to varying available bandwidth on the networks depending on changing capacity demands. Under normal conditions, however, a family can watch several television programmes concurrently in their home on different devices without experiencing any major perturbations.

3.2.3 Mobile Broadband Networks

With the advent of the first mobile networks in the 1980s communication changed fundamentally. Before that a phone call could only be made when using a telephone connected to a fixed land line. The availability of mobile phones revolutionized the habits of consumers. Getting used to the new liberty of communicating basically everywhere the desire for new capabilities beyond simple voice services grew. Coverage of mobile networks became significantly better over the years. However, the promise of ubiquitous coverage with always satisfying capacity and quality of service is hardly true even today. Nevertheless, mobile equipment manufacturers and network operators were pushing for the introduction of new features in mobile networks. Hence, several generations of mobile technologies have been developed and introduced around the world. The key differences between the different generations are summarized in Table 3.1 [GSM14].

The next generation of mobile network technology which for the moment just runs under the label "5G" shall be ready for deployment after 2020. This is believed to further push the limits and shall allow customers accessing any kind of service at any time in any location. More details on 5G and its potential implication on broadcast content distribution can be found in Sect. 6.3.

The introduction cycles between different generations of mobile network technologies are amazingly constant. Following [GSM14] every ten years a new generation of mobile technology has been introduced, starting with analogue mobile networks in 1980 and 4G (LTE) in 2010 while 5G is expected to be introduced around 2020. At the same time the peak market penetration of a mobile generation is reached about 20 years after its market launch. As a consequence, mobile network operators (MNO) have to maintain more and more different generations of network technology in parallel. While this entails huge investments it is far from clear where the revenues will be coming from. Whether such a development is sustainable for mobile network operators from a business point of view remains to be seen.

Table 3.1 Overview of mobile technology generations, their primary service offers and key features

Generation	Primary service	Key feature
1G	Analogue phone calls	Mobility
2G	Digital phone calls	Mass adoption
3G	Phone calls, messaging	Internet access
3.5G	Phone calls, messaging, broadband data	Broadband Internet, Apps
4G	All-IP	Faster broadband, lower latency

A mobile network is on one side essentially a terrestrial network. This means that in order to provide service within a given area a number of base stations are deployed. Mobile phones connect to one of the base stations from which they receive the services users are requesting. Similar to network planning for terrestrial broadcasting each base station will cover a certain area. For planning purposes mobile network coverage is typically modeled in terms of hexagonal coverage areas attributed to each base station. Coverage areas of base station are usually called a mobile cell. If a user is moving in a car or public transport, hand-over between base stations is to ensure that there is no loss of service when crossing from one cell into an adjacent one.

However, the visible terrestrial network is just one part of the mobile network. Data packets have to be carried to and from the base stations to those servers which actually provide the requested service such as a phone call, a text message, or Internet access. In contrast to fixed broadband networks, mobile network operators use either their own fixed network infrastructure or lease capacity from other network operators including fixed link communications to connect their base stations.

Even though the connection of the base stations to the backbone network is usually very powerful there are bottlenecks for the overall throughput of a base station. This holds in particular for situations where there are many concurrent users logged in to the same base station. As every base station can only offer a certain maximum traffic throughput depending, for example, on the available spectrum and the type of requested services it is quite common that unicast requests can only be served with significant time delay or service quality reduction. Therefore, service degradations can often not be avoided.

The cost issue for distribution of broadcast content over mobile broadband networks is quite similar for broadcasting companies as it is in the case of fixed networks. More traffic generates more costs. Broadcasters have to cope with this one way or the other. However, on the side of the consumers the situation is different. Typical contracts with a mobile network operator allow for a maximum amount of data per month at a speed high enough to be able to consume streamed audio-visual content. Once this limit is reached, the speed is throttled to a level which may still allow listening to radio programmes but TV watching becomes impossible.

Furthermore, the typical data caps coming along with mobile contracts may be 1GByte or 2GByte per month. If users are prepared to spend a lot of money, they may get 10GByte per month. Assuming that a video watched on a smartphone or a tablet in decent quality consumes 1MBit/s, one can easily calculate that even with the fantastic 10GByte contract only about 20h video watching per month would be possible. Having in mind that the average monthly watching time of linear TV in the USA in Q2 of 2015 still lay above 120h (see, for example, [Mar15]), it becomes clear that audio-visual content consumption through mobile broadband networks is still complementary to other forms.

3.2.4   WiFi Networks

Besides mobile broadband networks there exists another type of network which allows handheld and portable IP-enabled devices to connect to the Internet. These are WiFi networks. WiFi comprises a set of different standards which are issued by the Institute of Electrical and Electronics Engineers (IEEE). The base version of the standard was released in 1997, and has had subsequent amendments. With the latest version of the standard, i.e. IEEE 802.11af, data rates of up to 570 MBits/s can be reached [IEE16].

Even though WiFi is mostly perceived as a wireless extension to a fixed broadband network and hence not considered as an independent network, WiFi has become absolutely crucial for broadcasters in their effort to make available their content over broadband networks.

Wifi has a very attractive feature compared to mobile networks. While a mobile network always comes along with a mobile network operator to whom a user needs to subscribe, WiFi networks do not necessarily require a network operator. First of all, normally the scale of application is different. WiFi provides wireless Internet access within a limited area only while mobile broadband networks aim to cover whole countries or at least the populated parts of it.

WiFi uses unlicensed spectrum which anyone can use subject to the condition that there will be no regulatory protection for any service offered through WiFi. Anyone can set up his or her own WiFi network, for example, at home. The only thing which is needed is an Internet connection, for example, by means of DSL plus a WiFi enabled router.

As it is so easy and convenient to establish a WiFi access point to the Internet, usually a large number of WiFi networks are on-air in large apartment blocks or office buildings. In a limited geographical area this typically leads to increased interference levels. As there is no network management this may impair the user experience and therefore slow Internet connections.

There have been attempts in the past to offer WiFi coverage throughout larger areas, for example big cities like San Francisco in the USA [San07]. The purpose of the proposal was to provide free, wireless Internet access throughout San Francisco so that anyone with a computer and wireless access device could log in. Wireless access points would be mounted on light poles throughout the city to provide coverage. Even though from a user point of view this would have been a very attractive opportunity the project failed due to lack of financial support.

Each smartphone today is WiFi-enabled, thereby allowing users to switch from mobile network access to WiFi in order to contain their data consumption and cope with data caps which come as part of their mobile subscriptions. In particular, consumption of audio-visual content drives data usage. Therefore, as soon as there is a WiFi network available, for example at home, users are accessing audio-visual content including broadcast content on smartphones and tablets by connecting to WiFi and not through mobile broadband networks (see for example [Ofc15]).

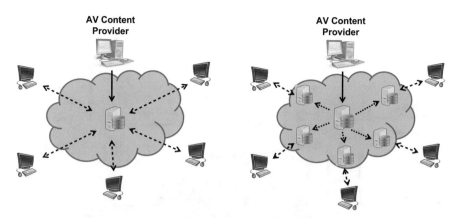

Fig. 3.6 Comparison between central server and CDN approach for the distribution of audio-visual content

3.2.5   Content Distribution Networks

A content delivery network or content distribution network (CDN) is a system of specifically distributed and configured servers which allows providing, for example, audio-visual content to users in a more reliable way and with higher performance than typical best-effort networks can do. It consists of a system of geographically distributed cache servers which contain copies of content [Alc11].

The standard way of hosting content on a webserver is to use a single central server which has to serve all users. In a CDN users are accessing different servers in search of the same content which helps to avoid traffic bottlenecks leading to service degradation and to optimize bandwidth usage. Figure 3.6 sketches the difference between the central server and the CDN approach for the distribution of audio-visual online content.

In reality, the topology of a CDN is more complex than depicted in the high-level overview of Fig. 3.6. In CDNs several basic types of servers can be distinguished. Any user requesting some content is served from a so-called delivery node. These need to be deployed as close to the edge of the CDN, i.e. near to the consumers, as possible. They host copies of all forms of data which are offered to the users.

Delivery nodes are supported by storage nodes whose primary purpose is to provide the required data to the delivery nodes. Origin nodes contain the original content from the content provider. They can make part of the CDN infrastructure or may still be located within the network of the broadcasting company, for example. The operation of the CDN is monitored and steered by the control nodes. They keep an eye on the communication between all nodes in the CDN and monitor the functionality of all CDN components. Figure 3.7 gives a simple overview.

CDN nodes are usually deployed in multiple locations, often distributed over multiple backbones. The number of nodes and servers making up a CDN varies, depending on the architecture. Some CDNs reach thousands of nodes.

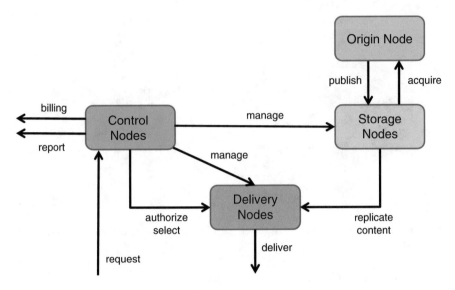

Fig. 3.7 Overview about the different CDN components

Users do not connect directly to a CDN. Instead, CDN nodes are connected to the network of an ISP. A user request for content is routed to a CDN by an ISP the user is subscribed to. This means for users the process is completely transparent as they see no difference in how they access content, whether they are being served by a central server or a CDN node.

Requests for content are directed to delivery nodes that are optimal with respect to resource allocation and costs. When optimizing for performance, locations that are best for serving content to the user may be chosen. This may be measured by choosing locations that are the fewest hops, i.e. the least number of network seconds away from the requesting client, or the highest availability in terms of server performance (both current and historical) in order to optimize delivery across local networks. When optimizing for cost, locations that are least expensive may be chosen instead [CDN16].

CDNs are not only employed for the delivery of audio-visual services but serve a large fraction of the Internet content today, including any type of web objects, downloadable objects, applications, and social networks. But, clearly, the advantages of CDNs become more and more pronounced the larger the amounts of data packages become which are requested by users. Naturally, TV content requires such high data rates. With the introduction of higher quality TV programmes on broadband networks such as HDTV or UHDTV, the application of CDNs becomes a vital issue. Without CDNs it will probably not be possible to make available broadcast content in an online world in a sustainable way.

Most broadcasting companies have contracts with CDN providers to carry broadcast content over their CDN networks to the users. In turn, CDN providers pay ISPs, carriers, and network operators for hosting their servers in corresponding data centers. Some broadcasters indeed operate their own CDN infrastructure some of which is also supporting the cooperation of distributed production studios rather than offering content to users.

Chapter 4
Regulatory Framework

Any distribution of broadcast content is subject to regulation. The details of the regulatory framework may vary from country to country. National administrations have developed their own rules thereby putting in force their corresponding national political directives. Europe is special in that respect as there is also regulation on a European level applicable throughout the European Union for issues relevant to Europe as a whole. This constitutes binding law for European Member States. However, according to the EU rules, national regulation must be brought in line with the EU law in order to avoid conflicts.

There are two aspects of regulation which are of primary relevance for the discussion here. As distribution of broadcast services utilizes some kind of telecommunication system a frame for the usage of appropriate electromagnetic spectrum is required. This is obvious in the case of terrestrial or satellite networks where an operator uses a certain range of spectrum to transmit a signal carrying the audio-visual content. Here the term terrestrial is to be understood literally thus including broadcasting and mobile networks. Even on cable networks or for Internet communication certain frequencies are reserved to transmit dedicated signals.

Spectrum regulation deals with assigning parts of the electromagnetic spectrum for dedicated usage. Since usually there are many users using the same or adjacent parts of the spectrum one of the major issues of spectrum regulation is to ensure compatibility between services by establishing corresponding sharing rules. There are several levels of spectrum regulation for which corresponding global, regional, or national organizations or bodies are responsible (see Sect. 4.1).

The second important aspect is content regulation. With regard to audio-visual content this defines what type of content can (or shall) be provided to whom, by whom, and under which condition. This also can include licensing rules, if any, for the right to provide AV content to the public.

Before the advent of the Internet and the possibility to deliver audio-visual content also by means of broadband networks the issue was well defined. Broadcast networks were the only networks which could offer television and radio

© Springer International Publishing Switzerland 2017
R. Beutler, *Evolution of Broadcast Content Distribution*,
DOI 10.1007/978-3-319-45973-8_4

programmes. Therefore, operation of broadcasting networks, in particular terrestrial broadcasting networks, was subject to double regulation, i.e. spectrum and content regulation. For telecommunication networks only spectrum regulation applied. There is an ongoing difficult discussion about this as these rules no longer seem to properly reflect reality in an Internet dominated world.

4.1 Spectrum Regulation

Telecommunication is omnipresent today. Nearly every household in the western world owns several radio receivers and most of households have at least one television set. Mobile phones are widespread and some people possess more than one phone. In some countries, for example in Scandinavia, already in 2003 the number of customers having exclusively a mobile phone contract exceeded the number of fixed line subscriptions. Nowadays, further wireless telecommunication systems are pushing into the market trying to gain ground and customers.

However, there are also many other services and applications that use spectrum beyond broadcasting and mobile services. These are satellite services, navigation and transport, maritime communication, public safety services, short range devices (e.g., to open/close cars), defense applications, and scientific usages. All these systems are utilizing a part of the electromagnetic spectrum. Spectrum regulators have a task to balance all requirements and ensure an efficient use of the spectrum.

In order to guarantee interference free coexistence between different telecommunication systems spectrum usage has to be organized. To this end, the electromagnetic spectrum is subdivided into many bands in which particular services such as broadcasting, mobile or fixed services can be operated under certain conditions.

The usage of any kind of electromagnetic spectrum is strongly regulated. There are three different levels of regulation, i.e. international or global, regional, and national regulation. On each level there are organizations which set up the rules for spectrum usage.

4.1.1 International Telecommunications Union

Spectrum regulation on an international or global level is taken care of by the International Telecommunications Union (ITU) [ITU16] which has been founded in 1865. It constitutes one branch of the United Nations (UN) and is subject to UN's rules of procedure. The ITU was created to act as an impartial international organization giving a framework in which national governments represented by their administrations and industries can work together in order to lay down the rules under which they can all successfully operate telecommunication networks and provide services.

As a matter of course, everyday people around the globe use their telephones to talk to each other. In the past this was done mainly via fixed telephone connections, in the meantime mobile phones are dominating. Access to the Internet, sending and receiving an email has become irreplaceable both in the business sector and in private life. Also, travelling is more and more depending on telecommunication services. This refers to planning business or leisure trips via the Internet or relying on navigation systems based on GPS. Furthermore, short range devices are penetrating our daily lives which means, for example, that any new car which is sold somewhere on the planet is equipped with corresponding devices to open and close the doors. All these examples make use of some kind of telecommunication system which benefits from the work of the ITU of managing the electromagnetic spectrum.

The ITU maintains and tries to extend international cooperation between each of its Member States, in order to allow for a rational use of any kind of telecommunication systems. Organizations and companies in the field of telecommunication are encouraged to participate in all corresponding activities, i.e. research, development, and standardization. One of the main objectives of the ITU is to offer technical assistance to developing countries in terms of mobilizing any kind of resources to improve access to telecommunication services in these countries.

The structure of the ITU reflects its main tasks. It is subdivided into three sectors, namely Radiocommunication (ITU-R), Telecommunication Standardization (ITU-T), and Telecommunication Development (ITU-D). Their activities cover all aspects of telecommunication, from setting standards to improving telecommunication infrastructure in the developing world. Each of the three ITU Sectors works through conferences and meetings, where members negotiate agreements which serve as the basis for the operation of global telecommunication services. Study groups made up of experts drawn from leading telecommunication organizations worldwide carry out the technical work of the Union, preparing the detailed studies that lead to authoritative ITU Recommendations.

ITU-R establishes technical characteristics of terrestrial and space-based wireless services and systems. It also undertakes important technical studies which serve as a basis for the regulatory decisions made at radiocommunication conferences. In ITU-T experts prepare the technical specifications for telecommunication systems, networks and services, including their operation, performance, and maintenance. Their work also covers the tariff principles and accounting methods used to provide international service. ITU-D experts focus their work on the preparation of recommendations, opinions, guidelines, handbooks, manuals, and reports, which provide decision-makers in developing countries with "best business practices" relating to a variety of issues ranging from development strategies and policies to network management. Each Sector also has its own Bureau which ensures the implementation of the Sector's work plan and coordinates activities on a day-to-day basis.

ITU-R is the sector that is tasked to manage the usage of the electromagnetic spectrum and satellite orbits. The technical characteristics and operational procedures under which the spectrum can be utilized are developed here. Member States develop and adopt a large set of particular rules for spectrum usage, called the Radio Regulations (RR) [ITU15]. They serve as a binding international treaty governing the use of the radio spectrum by some 40 different services around the world. Since the global use and management of frequencies requires a high level of international cooperation, one of the principal tasks of ITU-R is to oversee and facilitate the complex inter-governmental negotiations needed to develop legally binding agreements between sovereign states. These agreements are embodied in the RR and in regional frequency plans. The RR apply to frequencies ranging from 8.3 kHz to 3000 GHz, and now incorporate over 1000 pages of information describing how the spectrum may be used and shared around the globe.

An important component of the RR is the so-called Table of Frequency Allocation (TFA) in Article §5 of the RR. It describes in detail which part of the electromagnetic spectrum can be used in which geographical region by which service and under which conditions. The portion of the spectrum suitable for telecommunications is divided into "blocks," the size of which varies according to individual services and their requirements. These blocks are called "frequency bands," and are allocated to services on an exclusive or shared basis. The full list of services and frequency bands allocated in different regions forms the TFA. Even though a particular frequency band might be allocated to a special service like broadcasting, mobile or fixed service, this can be overruled or extended by means of footnotes containing special arrangements between individual countries.

Changes to the Table of Frequency Allocations and to the Radio Regulations themselves can only be made by a World Radiocommunication Conference (WRC). Modifications and revisions of the RR are achieved on the basis of negotiations between national delegations, which work to reconcile demands for greater capacity and new services with the need to protect existing services. If a country or group of countries wishes a frequency band to be used for a purpose other than the one listed in the TFA, changes may be made provided a consensus is obtained from other Member States. In such a case, the change may be indicated by a footnote, or authorized by the application of an RR procedure under which the parties concerned must formally seek the agreement of any other nations affected by the change before any new use of the band can begin.

In addition to managing the TFA, a WRC may also adopt assignment plans or allotment plans for services where transmission and reception are not necessarily restricted to a particular country or territory. In the case of assignment plans, frequencies are allocated on the basis of requirements expressed by each country for each station within a given service, while in the case of allotment plans, each country is allotted frequencies to be used by a given service, which the national authorities then assign to the relevant stations within that service.

With the help of its Bureau, ITU-R acts as the central registrar of international frequency use, recording and maintaining the Master International Frequency Register (MIFR). It contains entries for more than a million terrestrial frequency assignments and more than 100000 entries relating to different satellite services.

ITU-R also maintains broadcasting and satellite frequency plans (e.g., GE06 [ITU06a]). This is closely related to the fact that ITU-R is the central organization coordinating efforts that ensure that all the different telecommunication services can coexist without causing harmful interference to each other. Several computer based tools are offered to Member States to carry out corresponding analyses.

ITU-R prepares the technical groundwork which enables radiocommunication conferences to make sound decisions, developing regulatory procedures and examining technical issues, planning parameters and sharing criteria with other services in order to calculate the risk of harmful interference.

4.1.2 Regional Organizations

Different regions of the world have different requirements in terms of spectrum usage. Therefore, several regional organizations have been created representing groups of countries to develop and enforce their common view at an ITU level. Regional organizations also have a role at WRCs where they facilitate negotiations between regions, both before and during conferences.

CEPT, CITEL and APT are introduced briefly in the following. Information about the African Telecommunications Union (ATU) [ATU16], the Regional Commonwealth in the Field of Communication (RCC) [RCC16], and the Arab Spectrum Management Group (ASMG) [ASM16] can be found on their respective websites.

4.1.2.1 European Conference of Postal and Telecommunications Administrations

The European Conference of Postal and Telecommunications Administrations (CEPT)[1] [CEP16] is a European regional organization dealing with postal and telecommunications issues and presently has members from 48 countries. It was founded in 1959. Its basic objective is to deepen the relations between members, promote their cooperation, and contribute to the creation of a dynamic market in the field of European posts and electronic communications. Any European country can become a Member of CEPT as long as it is a member of the Universal Postal Union (UPU) [UPU16] or a Member State of the ITU. In 1988 CEPT decided to create ETSI [ETS16], the European Telecommunications Standards Institute, into which all its telecommunication standardization activities were transferred.

Representatives of ITU and UPU are usually invited to assemblies of CEPT while other inter-governmental organizations may be invited to participate as observers. This is also possible for organizations having signed a memorandum of understanding with CEPT declaring to subscribe to the rules of procedure of CEPT. Finally, the European Commission and the Secretariat of the European Free Trade Association are invited to participate in CEPT activities in an advisory manner, with the right to speak but not to vote.

---

[1]The acronym "CEPT" stems from the French name "Conference Européenne des Administration des Postes et des Télécommunications."

The CEPT has a hierarchic structure consisting of several bodies collaborating in clearly defined way. The highest body of CEPT is the Assembly which is chaired by the Presidency. The latter also acts a the secretariat for the Assembly which adopts major policy and strategic decisions and recommendations within the postal and electronic communications sectors. Committees may be set up by the Assembly dealing with different tasks assigned to them. Currently (2016) there are three committees, namely the European Committee on Postal Regulation/Comité Européen de Réglementation Postale (CERP), the Electronic Communications Committee (ECC) responsible for radiocommunications and telecommunications and the Committee for ITU-Policy (Com-ITU). Com-ITU shall organize the coordination of CEPT actions between CEPT administrations in the preparation for important ITU events, i.e., meetings of the ITU Council, Plenipotentiary Conferences of ITU, World Telecommunication Development Conferences and World Telecommunication Standardization Assemblies.

Each committee has several Working Groups dedicated to special aspects of postal and telecommunications issues. The Committees and the Working Groups are supported by the European Communications Office (ECO) located in Copenhagen. ECO is the distribution point for all ECC documentation and also provides detailed information about the work of the ECC and its Working Groups via the ECO web site [ECO16]. The major responsibilities of ECO include the drafting of long-term plans for future use of the radio frequency spectrum at a European level. Furthermore, national frequency management authorities of CEPT Members are supported in their work. Consultations on specific topics or the usage of parts of the frequency spectrum are conducted by ECO. An important task is also the publication of ECC Decisions and Recommendations and to maintain the record of the implementation of ECC Recommendations and Decisions in CEPT countries. An overview about the current CEPT structure can be found in [CEP16] while CEPT deliverables are publicly available at [CEP16a].

Three of the Working Groups of ECC have outstanding meaning for the usage of the electromagnetic spectrum in Europe. The first one is the Working Group Frequency Management (WGFM), the second is called Working Group Spectrum Engineering (WGSE), and finally there is the Conference Preparatory Group (CPG). WGFM is covering all issues which are connected to allocating spectrum to different telecommunication services like the harmonization of spectrum for Short Range Devices (SRD), Public Protection and Disaster Relief (PPDR), Programme Making and Special Events (PMSE), communications onboard vessels or aircrafts, spectrum for drones and others. Broadband Fixed Wireless Access in the 3.5 GHz and 5.8 GHz bands is another issue as well as any activities connected to the harmonization of frequency bands for the mobile service. Spectrum issues in relation to broadcasting are also addressed by WGFM.

In WGSE more technical issues of spectrum usage are dealt with. In particular this refers to the preparation of technical guidelines for the use of the frequency spectrum by various radiocommunication services. Furthermore, sharing criteria between radiocommunication services, systems or applications using the same frequency bands are studied and assessed by WGSE. This is directly connected to the investigation of compatibility criteria between radiocommunication services using different frequency bands. Results of studies and discussions are to be published in terms of CEPT-Recommendations and CEPT-Reports as necessary. The preparation of draft ECC Decisions lies also in the scope of WGSE as far as purely technical issues are to be addressed. WGSE contributes to the preparation of WRCs of the ITU and also to the work of Study Groups of ITU-R. Collaboration with ETSI and other international and European organizations is part of WGSE's task, too.

The third Working Group of CEPT having significant relevance for spectrum usage issues is CPG. From a strategic point of view it might even be considered the most important one because its scope comprises in particular the preparation of agreed European positions to be forwarded to ITU conferences like WRCs or Radiocommunication Assemblies (RA). Moreover, CPG is to develop, as required, coordinated positions in order to assist CEPT Administrations which are Members of the ITU Council in presenting a European position in respect of discussions concerning Conference agendas and timing. Furthermore, CPG organizes the preparation of the so-called European Common Proposals (ECP) which are submitted to WRCs as European contributions. These positions are supported by explanatory documents, called Briefs which provide necessary background information.

The CEPT maintains the so-called ECO Frequency Information System (EFIS) [EFI16]. EFIS is the tool to provide information regarding spectrum use in Europe. Basically, this is an excerpt from the RR including all relevant footnotes. It was cast into the form of a web interface. It allows easy and straightforward access to all information about the spectrum usage in Europe within a given frequency band. EFIS also contains national frequency allocation tables and the associated regulatory documents.

4.1.2.2 Inter-American Telecommunication Commission

The American counterpart to CEPT is called Comisión Interamericana de Tele-comunicaciones (CITEL) or the Inter-American Telecommunication Commission [CIT16]. Members are administrations of states of the Americas. However, the organization is open to any entity, organization, or institution related to the telecommunications industry to participate as an associated member. International or regional telecommunication organizations such as ITU can become associated members as well.

The primary objective of CITEL is to facilitate and promote the development of telecommunications in the Americas. Most of CITEL's work is done within the framework of committees that meet periodically and also coordinate their activities. Until 2018 the CITEL structure consists of an Assembly with an associated Steering group. Both are supported by the secretariat of CITEL. There are three committees, namely the Permanent Executive Committee (COM/CITEL), the Permanent Consultative Committee I (PCC.I) for "Telecommunications/Information and Communication Technologies (ICT)," and the Permanent Consultative Committee II (PCC.II) dealing with "Radiocommunication including Broadcasting." All of them can create working groups in order to fulfill their tasks.

COM/CITEL is the executive organ of CITEL. It is composed of representatives of thirteen Member States of CITEL elected at the Regular Meeting of the Assembly of CITEL. It has two major objectives, i.e. to carry out preparatory work for meetings of the ITU Council and world conferences and to develop the strategic plan of CITEL.

PCC.I constitutes an advisory committee of CITEL especially with respect to telecommunication/ICT policy, regulatory aspects, standardization, universal service, economic and social development, environment and climate change, and the development of infrastructure and new technologies. All relevant developments shall be monitored and discussed in order to harmonize the deployment of telecommunication services in the Americas. Currently, there are three working groups on regulation, development of technologies, and deployment of technologies.

Similarly, PCC.II acts as a committee covering frequency planning issues, coordination between Member States, harmonization and efficient usage of spectrum. Radiocommunication services and broadcasting across different distribution platforms are in the scope of PCC.II. Under this committee there are five working groups covering spectrum usage issues for different services such as satellite, mobile, or broadcasting services. Therefore, the committees PCC.I and PCC.II can be considered as CITEL's corresponding groups to the CEPT working groups WGSE and WGFM.

CITEL offers several databases providing information about spectrum usage in the Americas. In particular, from the CITEL website information can be obtained about regional mobile services as well as allocation of spectrum to services can be queried [CIT16].

4.1.2.3 Asia-Pacific Telecommunity

The role that CEPT and CITEL play in Europe and the Americas is filled by the Asia-Pacific Telecommunity (APT) in Asia [APT16]. APT was founded in 1979 and since then it brings together administrations, telecommunication services and network providers, manufacturers and R&D organizations to foster the development and deployment of telecommunications in Asia and the Pacific region. The preparation of global ITU conferences such as the ITU Plenipotentiary Conference (PP) or the World Radiocommunication Conference is within the scope of APT as an important objective.

APT is promoting development and regional cooperation in telecommunications, including radio communications and standard development. It undertakes studies relating to developments in telecommunication and information infrastructure technology. APT seeks to facilitate coordination between administrations within the region with regard to deploying telecommunication services.

The primary objective of APT is to support the development of telecommunication services and information infrastructure throughout the Asia-Pacific region. Particular emphasis is given to the expansion and development of telecommunication services in less developed areas in Asian and Pacific countries. APT extends from Iran in the West over India, China to Japan in the East. All of Indonesia and the Philippines is included as well as Australia and New Zealand.

APT has a structure which consists in the first place of a General Assembly, a Management Committee and a Secretariat. The Management Committee has the task to implement the policies and decisions of the General Assembly, while the Secretariat has a similar function like ECO in the case of CEPT.

The working areas of APT are Policy and Regulation, Radiocommunication, Standardization, Human Resources, and ICT Development. This structure is largely along the lines the ITU is organized which facilitates contributing to the ITU work. APT publishes several documents for its members and the public. Those are the annual APT Yearbook, an APT e-Newsletter, and also reports on workshops or studies carried out under the custody of APT. All this and more can be found on the website of APT [APT16].

4.1.3 National Regulatory Authorities

On a national level each country maintains one or more regulatory authorities which are to implement the national decisions about spectrum policy. Clearly, these implementations have to be in line with decisions taken on a regional or international level. Usually, there are different national authorities responsible for different kind of spectrum use (e.g., telecommunications, broadcasting, defense). In addition, in most countries spectrum and content are regulated by different authorities.

A national administration is free to use the spectrum for whatever purposes they wish as long as, for example, they do not cause interference to service in neighboring countries or claim protection from these services. The Well-known regulatory authorities are, for example, Bundesnetzagentur in Germany [Bun16], the Office of Communications (Ofcom) in the UK [Ofc16], or the Federal Communications Commission (FCC) in the USA [FCC16].

National regulatory authorities are actively participating in the corresponding regional organizations as well as in ITU activities. In particular, at an ITU level there are groups and meetings in which only national administrations are accepted to act through their national regulatory agencies. Those are, for example, the Plenipotentiary Conference or a WRC.

Within a country spectrum usage is governed by the rules of the national regulator. For the case of terrestrial broadcasting for example, regulators issue licenses to network operators which give them the right to use a particular frequency range to offer a broadcast service such as television or radio programmes. Typically, the licenses have limited duration ranging between 5 to 15 years and are subject to technical conditions for the network operation. These conditions are usually the result of bi- or multilateral coordination activities with those neighboring countries which may be affected by the operation of a transmitter using a particular frequency.

Conditions for spectrum usage come with very detailed descriptions of the allowed technical parameters of a transmitting station. In the first place, the maximum transmitting power is specified together with an antenna pattern which has to be implemented in order not to cause interference to other services. Furthermore, the height above ground where the antenna is mounted at a mast is another important technical parameter which determines the reach of the transmission of an electromagnetic signal.

4.1.4   European Commission

The European Commission (EC) plays a special role in Europe when it comes to spectrum regulation. On one hand, EC does not have the power to act as an independent player on a regional or international level because spectrum regulation lies with the sovereignty of the European Member States. This means the EC cannot represent Member States in terms of spectrum policy or regulation at the ITU or CEPT. As a consequence, in international meetings such as a WRC the EC has only the status of an observer. Therefore, they cannot make an official contribution to the Conference, nor take the floor and speak on behalf of the European Member States.

On the other hand, the EC coordinates and determines European policy in many fields in particular those which have an economic dimension. This may lead to EC Decisions which become binding law for European Member States. Spectrum policy is considered one the most important fields which has a direct relation to economic development and prosperity in Europe as it touches upon the key market of telecommunication. A common European approach to spectrum policy issues with the target to harmonize spectrum usage across Europe lies at the core of the engagement of the European Commission.

The EC spectrum policy is based on four primary areas of activity [EuC16]. As the usage of electromagnetic spectrum for any kind of telecommunication service is a matter of international coordination the EC tries to identify needs for spectrum coordination at an EU level. Secondly, harmonization of spectrum usage in particular frequency bands is supported and initiated if necessary. Very importantly, in case there are different requirements among European Member States for spectrum usage the EC establishes priorities. And finally, the EC is setting a regulatory environment in Europe to enable easier and more flexible access by public and private users to spectrum resources. In practical terms, the EC enforces harmonization of spectrum usage wherever this is possible, it supports any measures to increase the efficient use of spectrum and improves the availability of information about the current use, future plans for use and availability of spectrum.

In 2002 a framework for spectrum policy in the European Union was launched for telecommunication services. This led to the Radio Spectrum Decision (RSD) [RSD02] which defines the policy and regulatory tools to ensure the coordination of policy approaches and harmonized conditions for the availability and efficient use of radio spectrum for the European Member States. The RSD allows the EC to adopt decisions targeting to harmonize technical conditions with regard to the availability and efficient use of spectrum. In this context, the EC may issue mandates to CEPT for the preparation of corresponding technical implementing measures [EuC16].

The Radio Spectrum Decision initiated the creation of two complementary bodies in order to provide the EC with the necessary tools to facilitate consultation among European Member States and to develop and support a European spectrum policy. Those are the Radio Spectrum Policy Group (RSPG) [RSP16] and the Radio Spectrum Committee (RSC) [RSC16].

The RSPG is a group of high-level national governmental experts to help the EC developing a general spectrum policy. Its members are representatives of Member States and the Commission. RSPG contributes to the development of the European spectrum policy by considering not only technical issues but also economic, political, cultural, strategic, health, and social aspects. Potentially conflicting needs of radio spectrum users also fall into the terms of reference of RSPG always with a view to ensuring that a fair, non-discriminatory and proportionate balance is achieved.

A major work tool of RSPG is the possibility to publish the so-called RSPG opinions dealing with particular spectrum related issues such as coordination of policy approaches, harmonization and efficient spectrum usage. Furthermore, RSPG issues public consultations on certain activities and aspects to collect the views of stakeholders in Europe.

The second group, i.e. the Radio Spectrum Committee, is a committee set up to assist the EC in developing technical implementation measures to ensure harmonized conditions across Europe for the availability and efficient use of radio spectrum. The RSC is composed of Member State representatives and chaired by the European Commission. In contrast to the RSPG the work of RSC is more technology oriented. To this end, RSC is in the position to issue mandates to CEPT asking for the development of technical implementing measures. Decisions of the EC as far they relate to the technical aspects of spectrum usage are usually based on the results of investigations triggered by these mandates to CEPT. The majority of RSC documents are openly available to interested parties and the public [RSC16a].

4.2   Content Regulation

In contrast to how the usage of spectrum is organized, content regulation is primarily a national issue. Terrestrial broadcasting is a good example to this. At an ITU level the range 470–694 MHz is allocated to the broadcasting service on a primary basis in ITU-R Region 1. The actual usage of this spectrum range is subject to the provisions of the GE06 Agreement [ITU06a]. Along a common border application of GE06 may entail bilateral coordination between neighboring countries. As a result, each administration obtains the right to use a frequency at a given transmitter site under certain conditions in order to provide, for example, a terrestrial television service.

However, the question which programmes are offered is independent from this and rests entirely with the national regulatory authorities dealing with content regulation. They determine the number of programmes in a DTT multiplex and the envisaged coverage area and service quality. Furthermore, the national authorities decide independently what kind of programmes shall be offered. It may be public service broadcast content, content from commercial advertisement based broadcasters or even pay-TV resting on dedicated subscription schemes.

In the days of analogue television and radio services the regulatory environment was very simple. There was only linear broadcast content and it was delivered over one or the other broadcasting network. This could have been a terrestrial broadcasting network, a satellite, or even a cable network. In order to have access to content and receive the programmes the users had to purchase a dedicated broadcasting receiver which was usually attached to a roof-top antenna. Fixed reception was prevailing for television watching, while for radio portable and mobile reception has always been a very important use case.

All analogue broadcasting distribution networks can typically offer only limited transmission capacity. This has been significantly extended with the introduction of digital transmission technologies but still the capacity of the systems is limited. Therefore, regulators had to put measures in place which allowed the usage of the available transmission capacity in a way which was compliant with national media policy. In some countries such as Germany this led to a system of sharing on a 50 % basis between public service and commercial broadcasters, i.e. 50 % of available TV channels were given to public broadcasters and 50 % to commercial broadcasters.

The fact that a broadcasting receiver was needed to access broadcast content was used to decide which service would need to adhere to broadcast content regulation. With the advent of the Internet and the fact that through IP-based broadband networks also broadcast content can be accessed, difficult discussions were triggered on how content regulation has to be adapted in order to cope with these new developments. These discussions are far from being settled. Since the developments regarding offer and distribution of audio-visual content over the broadband networks are rather gathering momentum than slowing down, they will continue.

With the proliferation of other means to access audio-visual content than traditional broadcast distribution technologies the boundaries between different types of audio-visual services are getting blurred. Consequently, the regulatory regime needs to be adapted to be able to cope with these new challenges. This is important as audio-visual services have started to play a fundamental role in democratic societies. They are informing citizens, shaping the public opinion and offering a window to the world. This not only refers to audio-visual services provided by broadcasting companies. More and more user generated content is produced, the most prominent example being certainly YouTube, which enriches and helps the lives of citizens. All this content has to be made available and distributed across corresponding networks. In order to enable equal and transparent treatment of all services national administrations are endeavoring to develop corresponding regulatory rules.

4.2.1 The European Regulatory Framework

In Europe the European Commission is supporting a regulatory framework for audio-visual services which shall strengthen the European position from an economic point of view in the first place. Audio-visual services generate substantial economic value. They represent a huge industry from production and distribution to the device market. However, audio-visual content is also of particular importance because it relates to particular social values, cultural and linguistic identities. Therefore, the impact of audio-visual content reaches beyond commercial categories.

As audio-visual services relate to many different areas they are therefore subject to different regulation. These include sectors and issues such as telecommunication, copyright, competition, public service media, consumer protection, privacy and data protection; and net neutrality.

The EU's regulatory framework for electronic communications is a series of rules which apply throughout the EU Member States. They shall encourage competition, improve the functioning of the market, and guarantee basic user rights. The overall goal is for European consumers to be able to benefit from increased choice, thanks to low prices, high quality, and innovative services [EuC16a].

The European regulatory framework sets out five directives which represent binding law for EU Member States. Those are:

- the Framework Directive [EuP02] which establishes a harmonized framework for the regulation of electronic communications services, electronic communications networks, associated facilities and associated services, and certain aspects of terminal equipment to facilitate access for disabled users;
- the Access Directive [EuP02a] which harmonizes the way in which Member States regulate access to, and interconnection of, electronic communications networks and associated facilities;
- the Authorization Directive [EuP02b] which is to implement an internal market in electronic communications networks and services through the harmonization and simplification of authorization rules and conditions;
- the Universal Service Directive [EuP02c] which concerns the provision of electronic communications networks and services to end-users with the aim to ensure the availability of good-quality publicly available services throughout the European Union; and
- the Directive on Privacy and Electronic Communications [EuP02d] which provides for the harmonization of the national provisions required to ensure an equivalent level of protection of fundamental rights and freedoms, and in particular the right to privacy and confidentiality.

These Directives build the overarching framework of European regulation. Content carried over electronic communications networks including audio-visual content is regulated by the Audio-Visual Media Service Directive (AVMSD) [EuP10]. The AVMSD governs EU-wide coordination of national legislation on all audio-visual media; both traditional TV broadcasts and on-demand services.

The regulatory framework of the AVMDS spotlights several basic objectives such as

- to strengthen the single European market and to guarantee fair competition conditions;
- to support emerging audio-visual media services;
- to underpin the competitiveness of the European audio-visual industry and to promote European audio-visual content;
- to contribute to the support of cultural and linguistic diversity and heritage in Europe; and
- to safeguard media pluralism, freedom of expression and information.

This is to ensure key societal values and a high level of protection of objectives of general interest, for example the protection of minors and human dignity, promoting the rights of persons with disabilities and combating racial and religious hatred.

The AVMSD provides for a minimum harmonization of certain aspects of national legislation related to audio-visual media services. The Member States can apply more detailed or stricter rules to providers under their jurisdiction, as long as those rules are consistent with the general principles of EU law. Furthermore, the AVMSD is based on the principle of technological neutrality: rules apply to providers of audio-visual content or the delivery infrastructure irrespective of the screen on which the content is watched. In 2015 the AVMSD was under review as the audio-visual media landscape has changed significantly since the adoption of the Directive in 2007 [EuC15]. Corresponding changes are expected in 2017.

It is important to note that the AVMSD does not apply to content hosted by online video-sharing platforms, i.e. YouTube and similar platforms offering in particular user generated content. So far, these platforms are regulated by the EU eCommerce Directive [EuP00]. This is crucial because it means that for the time being this kind of regulation exempts online video-sharing platforms from liability for the content they transmit, store, or host, under certain conditions.

While the provisions on the Directive's scope have remained effective to date in Europe, this may no longer be the appropriate way forward in the future with increasing consumption of audio-visual services via broadband networks and through audio-visual platforms. Powerful video-on-demand (VOD) and OTT providers are emerging which are often active on a global scale. Their impact on and relevance is steadily growing. If they are not subject to the same regulatory framework as traditional broadcasting companies, a major imbalance would result.

Apart from an economic imbalance there may also be other issues, for example the protection of minors. TV content is strictly regulated in order to ensure that content not appropriate for minors are not offered on broadcasting networks, for example, between 6:00 in the morning and 22:00 in the evening. However, there are cases where TV programmes which are made available outside this corridor on broadcasting networks can be easily accessed all day through online video-sharing platforms, usually through third parties and not the original content providers themselves.

Advertising as the fundamental pillar on which commercial broadcasting rests is also regulated through the existing AVMSD. But also the majority of public service broadcasters depend on advertisement based revenue to supplement the public funding provided for the fulfillment of their remit. However, regulation of advertisement so far only applies to linear TV and radio programmes. The AVMSD lays down rules that regulate advertising from a quantitative point of view. For example, they set a maximum of 12 min of advertising per hour on television, define how often TV films, cinematographic works, and news programmes can be interrupted by advertisements and set the maximum duration of teleshopping windows.

The availability of bi-directional communication between content providers and users opens up a completely new opportunity of advertisement design. Personalized advertisement build around individual audio-visual media consumption is considered the Holy Grail of future audio-visual media business. By matching advertisement to the needs of users, media service providers are convinced this will open up new ways of marketing their programmes and associated content products. It is expected that this kind of tailored advertising will dominate in the future. However, personalized advertisement is closely linked to the protection of the right to privacy and confidentiality of users. Any future regulation of the audio-visual content environment applicable across all distribution paths has to properly take this into consideration.

The example just given regarding protection of minors on all platforms can be used to highlight another important issue not covered by the existing regulation. Third parties making available original broadcast content through their platforms and portals should not take any measures to modify this content or degrade the service quality. The integrity of broadcast content offered through independent platforms must be protected. Hence, platform operators should be required to respect signal and content integrity.

With the proliferation of broadband networks and Internet services it becomes absolutely crucial for broadcasters that their content remains findable. Many broadcast content offers are widely used through Internet access. This refers to streaming of linear TV and radio programmes as well as dedicated on-demand offers. However, it should be clear that the position of broadcasters in an Internet environment strongly depends on their popularity and their market share on traditional broadcasting networks. This is particularly important for public service broadcasters which have to fulfill their public remit.

European regulation tries to cope with this requirement by making the provisions which allow imposing must-carry rules. However, existing rules only cover linear broadcast services and broadcasting networks. Delivery over broadband networks is currently not covered by must-carry rules. Furthermore, must-carry-rules do not include all linear services but usually only a subset of the whole content offer. Must-carry rules are essential legal safeguards to ensure citizens' access to content of public service broadcasters and need to be preserved and updated in line with new technological and market developments. Their importance may increase in the future with regard to development of broadband content distribution.

As long as must-carry-rules only refer to broadcasting networks the issue may be more or less easy to deal with. However, with the increasing relevance of Internet platforms as a means of distribution of audio-visual content, including public service content, there is a need to think about appropriate platform regulation in order to cater for fair and balanced market conditions among different content distribution options and content providers. For the time being, it is an open question how to implement must-carry rules for public service broadcasting content in such an environment.

Beyond must-carry rules there are other issues related to audio-visual platforms which are critical from a public service broadcasting perspective. A prominent example is YouTube, which started as a video-sharing website for user-generated content but which nowadays increasingly includes semi-professional and professional audio-visual content. Broadcasting companies have realized that they have to exploit YouTube in order to offer their content but there is no possibility yet to influence YouTube such that public service content would be treated similar as on traditional broadcasting networks.

Another example which causes trouble to public service broadcasters is the way in which manufacturers of TV sets are trying to gain ground in the area of providing access to content through their own portals. The so-called connected TVs can access content on terrestrial, cable, or satellite networks while at the same time being connected to the Internet. In addition to linear content modern TVs offer a portal containing apps provided by third parties. The layout and the order of the apps is determined by the policy of the manufacturer. As there is no regulation with regard to these portals it becomes increasingly difficult for PSBs to remain visible under these circumstances.

4.2.2 Net Neutrality

Broadcasting distribution networks differ from broadband networks with regard to one very important aspect which has a regulatory implication. Broadcasting networks only carry broadcast content, i.e. TV and radio programmes, and the directly associated services such as electronic programmes guides (EPG). Hence, broadcasting networks are delivery networks for a clearly defined purpose. In contrast, broadband networks represent multi-purpose networks. In principle, any type of telecommunication service can be offered through broadband networks.

With the help of computers or IP-enabled portable devices users have access to a universe of different services ranging from email and websites to audio-visual content. As a matter of fact, these services require very different data rates. Even if the Internet access is very slow due to limited network resources it is usually not a problem to send or receive a simple email which contains only text. This is different when trying to access audio-visual content on the Internet. Depending on the available bandwidth and the workload of the networks through which content is transported to the users, the quality of services may be degraded up to no service being available at all.

It is natural that both from a service provider point of view and from the user side it may be a good idea to provision enough capacity on a broadband network which ensures delivery of certain services such as audio-visual content with a guaranteed quality. However, this is in conflict with the idea of the open Internet. Open Internet means that the full resources of the Internet are easily and equally accessible to all individuals and companies. The concept of the open Internet is closely related to open standards, transparency, lack of Internet censorship, and low barriers to entry [NeN16].

The opposite of the open Internet would correspond to a situation where companies or governments control usage of the Internet. They would determine who could use it, for which services, under what technical conditions, and at what costs. This may open door to a significant level of censorship leading to explicitly blocking unwanted content, either because of its political or cultural orientation, or because of the intention to protect national industries against foreign economic competitors.

The concept of net neutrality is often considered as the fundamental measure to achieve an open Internet. Net neutrality means to put forward basic policies in order to oblige Internet service providers and governments to treat all data on the Internet the same, not discriminating or charging differentially by user, content, site, platform, application, type of attached equipment, or mode of communication. As such, network neutrality represents an equal rights approach to using infrastructure resources to promote fair evolutionary competition in an economic environment [WuT03].

However, defining net neutrality only in terms of saying all data on the network shall be dealt with equally may give rise to conflicting interpretations. How can different services such a phone call, an email, or live streaming of a HD television programme properly be compared? This issue may be resolved by making reference to the famous OSI-model [OSI16] which deals with all traffic on the Internet by separating treatment of data into seven distinct hierarchical layers. A technical definition according to which a network operator would be in the position to implement net neutrality on his network could read:

> Network neutrality is the adherence to the paradigm that operation of a service at a certain layer is not influenced by any data other than the data interpreted at that layer, and in accordance with the protocol specification for that layer [BaE14].

Therefore, a network operator would operate his network in accordance with the principles of net neutrality if the network provides the service strictly in line with the specification of the network protocol it implements as its service. In other words, an ISP dealing with IP-packets at Layer 3 of the OSI-model shall perform this task without accounting for any other information beyond the Layer 3 interpretation of the network traffic, and without influence by any other logic than the Layer 3 networking specification [BaE14].

Blocking access to websites by an ISP would be a form of violation of net neutrality. This would be also true if an ISP would provide more bandwidth for particular services while throttling the data throughput for other services or would suppress the delivery of emails. The first example takes place on OSI-layer 3 only.

There an ISP could decide to block certain addresses. This corresponds to a violation of net neutrality where data of a certain OSI-layer is not used in accordance with the networking specification. The other examples correspond to an intervention where information obtained on OSI-layer 7 is used thereby violating the rule not to mix OSI-layer information [BaE14].

In a way, net neutrality is the opposite to the concept of managed service as discussed in Sect. 3.2.1. Managed services represent services for which a certain amount of bandwidth is reserved or for which a defined QoS is guaranteed. This can be either achieved by using a dedicated network infrastructure or reserving part of the available network capacity. Both ways are clearly in conflict with the idea of open Internet and net neutrality. The first example is trivial and the second requires some level of traffic monitoring across OSI-layers in order to prioritize some data over others.

For public service broadcasters the issue of net neutrality is both important and ambivalent. Net neutrality is a measure to achieve societal, cultural, and political goals to which PSBs have to adhere to. Net neutrality can provide open access to electronic communication for everyone without discrimination. The free exchange of views and opinions can be guaranteed as economic barriers to participate can be eliminated. This is very important for small start-ups which want start new businesses depending on unbarred and fast access to resources available on the Internet and the ability to exchange information with customers. Net neutrality can be seen as a motor for improving the Internet speed and stability for all communication. If ISPs can sell premium connectivity to certain players, it can be expected that those ISPs will no longer invest in general terms, but only for those connections that pay extra. As such, net neutrality is the guarantor for impartiality and transparency.

An open Internet based on net neutrality principles enabling indiscriminate access to audio-visual content of broadcasters basically corresponds to a best-effort network. Depending on circumstances such as time of the day, geographical area, and traffic demand it may be difficult or even impossible to watch a HDTV programme streamed over the Internet. This poses a severe problem to broadcasters as most of them have very strict regulatory obligations which govern, for example, also the level of quality of their services. Public service broadcasters have to guarantee reception of their TV and radio programmes almost 100 % of time and across large areas (see Sect. 7.1) which cannot be achieved by best-effort networks as of today.

Therefore, PSBs are also interested in having access to networks offering managed services. This may have significant economic implications. Reserving network capacity on a broadband network for the delivery of audio-visual services with guaranteed QoS independent of the number of concurrent users is an economic challenge for many PSBs. Nevertheless, broadcasters are looking to exploit both ways of using the Internet, i.e. open Internet and managed services, as otherwise they may lose ground against purely commercial content providers.

Net neutrality is a hot topic both in Europe and in the USA. The European Commission and the American regulator FCC have put forward concepts to deal with net neutrality. In October 2015 the European Parliament has adopted a set of rules regarding the implementation of net neutrality in the European Union [EuC15a]. It is considered as a vital measure to achieve the long-term goal of the Digital Single Market in Europe [EuC16b]. The European net neutrality rules constitute binding law for European Member States. They state that [EuC15a]

- every European must be able to have access to the open Internet and all content and service providers must be able to provide their services via a high-quality open Internet. From the entry into force of the rules, blocking and throttling the Internet will be illegal in the EU and users will be free to use their favorite apps no matter which offer they subscribe to. Many mobile providers are blocking Skype, Facetime or similar apps or sometime they ask extra money for allowing these services: this will be illegal.
- All traffic will be treated equally. This means, for example, that there can be no paid prioritization of traffic in the Internet access service. At the same time, equal treatment allows reasonable day-to-day traffic management according to justified technical requirements, and which must be independent of the origin or destination of the traffic and of any commercial considerations. Common rules on net neutrality mean that Internet access providers cannot pick winners or losers on the Internet, or decide which content and services are available.

It is important to understand that these provisions refer to the open Internet only. This means that there will be no paid traffic prioritization in the open Internet. However, it is recognized that there are different services in the Internet requiring different data rates and other technical parameters. Network management to balance all competing demands is still possible, however, not at the detriment of service providers or users. Any traffic management measures must be transparent, non-discriminatory, and proportionate.

In addition to the open Internet services the EU legislation allows for the so-called specialized or innovative service on condition that they do not harm the open Internet access. These are services like IPTV, high-definition video-conferencing or healthcare services like telesurgery for which the best-effort mechanism of the open Internet is not sufficient. However, such services must not be offered as a substitute for the open Internet; they may only come on top of it.

The European regulation sets clear and strong conditions for the provision of such specialized or innovative services [EuC15a]:

- They have to be optimized to provide specific content, applications or services.
- It must be objectively necessary to meet service requirements for specific levels of quality that cannot be met the provisions of the open Internet.
- These services can only be provided if there is sufficient network capacity to provide them in addition to any open Internet access service.
- They must not be to the detriment of the availability or general quality of the open Internet access for users.

The responsibility to monitor and enforce compliance with the net neutrality rules rests with national administrations and governments. This refers in particular to the issue of ensuring the availability and quality of the open Internet in case specialized or innovative services are offered. National regulatory authorities will have to make sure that the open Internet access is not affected or degraded by traffic discrimination through ISPs or by the provision of specialized services.

At a first glance, the new European regulation regarding net neutrality goes in the right direction from a public broadcaster's point of view. However, it will depend crucially on the active role of national regulatory authorities to implement and monitor the rules appropriately. Furthermore, it is important that network operators are fully transparent towards the authorities as well as towards consumers. In particular for public service broadcasting it is important that EU policy makers ensure that [EBU15]

- the open Internet is the norm and specialized services the exception;
- the development of specialized services must not impair open Internet access;
- there are clear transparency requirements to reinforce users' trust in the open Internet;
- clear rules make sure that equivalent types of traffic are treated equally;
- the specific cases in which network operators can manage Internet traffic are defined clearly; and
- undue content blocking or discrimination is prevented.

It is vital that broadcasters keep insisting on adhering to these principles as the immediate reaction of some of the German mobile network operators as a response to the adoption of the European net neutrality rules clearly showed. Deutsche Telekom supported by Vodafone declared that they would welcome the new regulation but at the same time indicated that those who are in need of high speeds or quality of service may have to pay in the future (see, for example, [EuA15]). This may give an impression that the debate about net neutrality in Europe is far from being settled and different interpretations of the new regulation will be put forward in the time to come.

The situation seems to be quite similar in the USA. In spring 2015 the national regulator FCC has put forward provisions to deal with net neutrality [FCC15]. These new rules are guided by three principles which are that the US broadband networks must be fast, fair, and open. The new regulation defines broadband Internet access services as mass-market retail services that allow access to the entire Internet, and it includes both wired (e.g., cable or fiber) and wireless (e.g., satellite or mobile) networks.

In essence, the regulation targets to protect free expression and innovation on the Internet and promote investment in the nation's broadband networks. The net neutrality rules seek to prevent Internet service providers from speeding up or slowing down some websites over others.

The new US regulation stipulates exceptions from the rules for specialized services which represent a loosely defined category of applications such as VoIP phone service or real-time health monitoring. Specialized services, though they might use Internet protocol and may travel over the same wires or airwaves as a broadband Internet access service, do not provide access to the Internet generally.

Actually, it lies with the ISPs to claim if a service falls into the specialized service category or not. This may give rise to serious issues, for example, if an ISP claims that its own video streaming service competing with a provider such as Netflix would be a specialized service thereby giving priority to his own service at Netflix's expense.

However, the regulatory framework permits FCC to include any service in the definition of broadband Internet access services if FCC finds that the service is providing the "functional equivalent" of broadband services. In other words, if it appears to the FCC that an ISP tries to evade the net neutrality rules by re-labeling a particular service as a specialized service the FCC has a right to intervene and apply net neutrality rules [Rua15].

Even though at a first glance the US rules seem to be more stringent, it still remains to be seen whether in practice they will be applied so rigorously.

Chapter 5
Changing Habits and Expectations

People are enjoying radio and TV programmes now for about a century. The world has changed fundamentally since the first radio broadcasts were launched. At the beginning the only way to receive radio and TV programmes was through analogue terrestrial broadcasting networks. In the meantime there are many other ways to access broadcast content, in particular through broadband networks (see Chap. 3). The switch-over from analogue to digital technology already opened up completely new opportunities but certainly the triumph of the Internet lifted distribution of broadcast content to another level.

Technology developments made huge steps in all areas of the broadcasting business. Not only distribution was revolutionized but also content production has reached new dimensions with the advent of digitization. Last but not least, the device sector is offering a universe of possibilities to receive, store, share, and consume broadcast content. All this led to a change of user habits which after some period of adaptation also generated new expectations.

This also had an impact on broadcast content providers which had to customize their offers to the new reality. New types of services emerged and in particular the direct interaction with the audiences through electronic communication means or social media has changed the business models of broadcasters for evermore. Who made the first step into this new world, i.e. was it users first embracing new technologies and broadcasters following or vice versa, is impossible to say and maybe also not relevant. The only thing that counts is that new opportunities have been generated which will determine any future developments.

The industry needs to understand where the technological development may go to in order to take proper strategic decisions now and keep their enterprises on a prosper track into the future. As a consequence, many people are analyzing the current situation in order to be able to predict the future. Clearly, there are many predictions which are in complete contradiction to each other. Apart from the difficulty to make reasonable forecasts, all these statements about the future are

© Springer International Publishing Switzerland 2017
R. Beutler, *Evolution of Broadcast Content Distribution*,
DOI 10.1007/978-3-319-45973-8_5

certainly biased by the corresponding expectations and needs of those carrying them out. This has to be borne in mind when looking at the examples given here.

The profound change which has taken place in recent years in audio-visual content production, distribution, and consumption can best be visualized when looking at three different areas separately, being well aware of the fact that all these changes and trends are closely linked and intertwined and mutually depend on each other. These three different areas are the services, i.e. the different types of content, the devices on which content is consumed and the conditions under which consumption takes place, i.e. at home, in a public place like an airport or a park and while being in motion, for example, in public transport or a car.

5.1 Extension of Services

Originally broadcasting companies were offering exclusively linear television and radio programmes. This means that whatever users could see on their television screens or could listen to on their radio receivers was broadcast at that moment. Clearly, the transmission was subject to tiny time delays due to processing of the signals and the time to travel to the users.

Linear broadcast service corresponds to a content offer where an editing department of a broadcasting company prepares a schedule of different types of programmes consisting of news, shows, soaps, dramas, movies, episodes of series, and so forth. The result is usually a 24h/7d programme. Users can tune in to this schedule of content; they may hop from one channel to another or completely switch off their receiver. However, they are not in a position to change the defined schedule of the TV or radio programme.

Live programming is a special variant of a linear service. Live transmissions are presenting an event contemporaneously as it happens. Well-known live programmes are quite often news, entertainment shows on a Saturday evening or sports events. They usually constitute only a part of the full linear programme offer over one day, for example the live broadcasting of a Champions League football match between 8pm and 11pm amended by all the pre- and post-reporting, interviews after the match and discussions. Once this is over the linear TV programme is continued with other content as part of the linear offer.

Linear broadcast content comes with the problem that if users want to see a particular programme or listen to a special radio show they have to switch on their receivers in time. Even though in former times it was quite common that families met in their living rooms to watch their favorite TV show at a given time, this is getting more and more difficult in western societies. Family life has changed significantly and many people do not want their daily lives be influenced by schedules of linear broadcast services anymore. Therefore, many people make use of different types of recording devices to store audio-visual content, for example, on local hard drives and then watch TV shows whenever they find the time to do so.

Indeed, this has become a widespread way of watching TV content thereby making users independent from the given schedule of the broadcasting companies.

This can be considered to fall under a category of new broadcast services which are called time-shifted services. However, broadcasting companies offer these kind of services also directly. The simplest way of implementing such time-shifted services is to offer, for example, the linear programme and copies of it which are shifted by 1 h, 2 h or more at the same time. Quite often all these programmes are bundled into a programme multiplex broadcast on a DTT network. When distributing linear content on broadband networks it is also possible to offer several streams of the same linear programme each of then shifted by a fixed time interval. Whether or not this is a reasonable use of network resources is something broadcasters have to decide themselves. These kinds of time-shifted content offers are probably not ideal solutions because even though they may give more freedom to the user there are still restrictions, i.e. the content provider decides about the time shifts and thus offers only limited choice.

Watching linear TV in such a static time-shifted manner will probably remain a niche application. What has certainly become very interesting to users is the possibility to watch a particular part of the overall linear programme at a later time defined by the user himself. This could be a music show, news, or very often an episode of a series which a user may have missed on linear programme. Therefore, most broadcasting companies offer parts of their content through dedicated Internet portals. Content which has been broadcast as part of the linear programme can be streamed after the linear broadcasting at any convenient time. This kind of service is usually called a catch-up service. Well-known examples are the iPlayer service of the BBC [iPl16] or the Mediatheken of the German public broadcasters ARD [ARD16] and ZDF [ZDF16]. In any case, catch-up services are not feasible on pure broadcast networks. They require access to a broadband network, at least to send the corresponding request to a server hosting the catch-up service.

Catch-up services may already be counted to a new type of broadcast service which could be introduced only with the new opportunities offered by the Internet. Clearly, this refers to the so-called nonlinear services. With linear services the control lies with the content provider. It is the broadcaster who decides which piece of content makes part of the content offer and at which time it is offered. With nonlinear services the control about consumption is transferred to the user.

This means that users are in a position to select individual pieces of content to consume them at their convenience. This can be a news show, entertainment, a movie, an episode of a series. Users can decide about the time of consumption, the location and the device on which consumption takes place. Clearly, this is all subject to technologies and infrastructures enabling this.

The crucial issue, though, is that the user has to become active. He or she need to scan the nonlinear offers until they find something they are interested in. Then, they have to make a corresponding request to the service provider before they can watch or listen to the audio-visual content.

Nonlinear content comprises many different forms. The two basic forms are streaming and downloading of audio-visual programmes. Both give the user the autonomy to decide when, where, and on which device a particular piece of content is used. In most cases, both forms of nonlinear content consumption require access to a broadband network. The difference between streaming and downloading is that once the request is sent for a stream the video or the audio starts assuming that the broadband connection is good enough, which is an issue to be addressed later. Furthermore, for a satisfying streaming experience the device has to be connected all the time. Downloading refers to storing content for later usage not necessarily on the device to which it was stored.

The content portfolio of broadcasting companies has been extended during the last ten years by making extensive use of social media. YouTube, Facebook, and Twitter have become integral and indispensable elements of every TV or radio show. YouTube started as an online platform for user generated content. Even though this is still the bulk of audio-visual content available on YouTube, broadcasters have realized that by creating dedicated YouTube channels they open up an additional distribution path for their content. There are YouTube channels referring to the entire programme or special channels featuring a particular element such as a sports or new show. ARD and BBC offerings are examples which can be easily found by going to the YouTube website and searching for ARD and BBC, respectively.

Establishing new distribution opportunities was only one reason why broadcasting companies jumped on social media, probably even a less important one. Far more exciting is the possibility to enter into a direct and instant dialogue with the audience. If a radio programme provides an associated Facebook page, listeners can easily contact the radio makers and express their wishes, give feedback, and comment on what they hear. Furthermore, they can contribute their own audio-visual content which relates to the current radio programme. This may range from pictures to videos about activities taking place while listening to radio.

For TV shows feedback coming through social media including simple emails has replaced the traditional call-in to get an audience poll during a life show. These are just some simple examples; in practice programme people are very creative to extend this kind of interactions in order to engage viewers and listeners. It also creates a strong bond between audience and a radio or TV station which is vital to survive in today's audio-visual content markets.

Offering independent linear and nonlinear services alongside with social media interaction is straightforward. However, the combination of all elements to create a holistic viewing or listening experience is one of the major issues for broadcasters today. Seamless switching from linear to nonlinear content and back is crucial. While watching a news show a user may become interested to get more information on a particular topic or may wish to recall something which has been broadcast some time ago on the same issue. Hybrid services is the label for these kinds of offers.

Technological solutions are available to enable hybrid services, some of which are based on apps. Broadcasters have engaged in the development of Hybrid Broadcast Broadband TV (HbbTV) [Hbb16]. HbbTV requires a TV which contains one or more traditional broadcasting receivers, i.e. terrestrial, cable or satellite, and at the same time can be connected to the Internet, either wired or wireless through WiFi. The HbbTV application allows easy and seamless access to the nonlinear offer associated with a given linear programme. This may be extra information for the current programme, an electronic programme guide or additional footage to be shown, which is not part of the linear programme. Whenever a linear service is HbbTV enhanced this is indicated to the user who by simply pressing the red button on the remote control is then led to the additional content offer.

HbbTV is the first step in the direction of creating completely new content types. However, broadcasters are experimenting with new formats which push the limits even further. The possibilities to combine linear with nonlinear content to offer a holistic content experience is transcended by what is called cross media production. This is the idea to exploit the particular characteristics of radio, TV, and the Internet in a complementary way. For example, when the script for a movie is drafted, a version of that movie is developed for radio and Internet at the same time.

Just doing this independently is actually nothing really exciting. The compelling new idea is that what is offered on TV, radio, and Internet is not the same. Rather, it is complementary. This means, for example, that on the Internet footage may be provided which is not part of the movie on TV, for example a scene shot from a different perspective such as not from the perspective of the main actor but rather another person. On radio there could be additional dialogues which can neither be found on TV nor on the Internet. The objective is to present audio-visual content such that the combination of TV, radio, and Internet adds up to a new enhanced experience beyond TV alone (see, for example, [SWR11]).

Developing and offering news services is definitely crucial for broadcasters. They need to safeguard their market position by adjusting the content portfolio to reflect current trends in audio-visual media consumptions. However, this requires carefully observing what users are actually doing or expecting they could be doing soon.

A very comprehensive overview about the markets and user behavior around the globe in the broadcasting sector can be found in a study which was carried out by IHS Technology for the World Intellectual Property Organization (WIPO) [WIP15]. IHS Technology is a globally active information company which is providing market intelligence for many different fields including media [IHS16].

The analysis of IHS Technology captures market data between 2004 and 2014 for many markets around the world. For six key markets, i.e. France, Spain, Germany, Italy, UK, and the USA they monitor the viewing times for linear television, time-shifted TV on the basis of recording to a hard drive, pay TV video-on-demand services and over-the-top (OTT) offers. They provide some very interesting findings:

- In 2014 television content, linear and time-shifted viewing, made up 96 % of all video consumption in the six countries investigated by IHS.
- There is a decline of linear TV consumption to be noted. It can be observed that the reduction is still very small lying between 1 % and 2 % per year. Nevertheless, linear TV remains very strong in each of the monitored markets with an average of 88 % of total viewing time and reaching as high as 94 % of viewing in some markets.

- With regard to nonlinear TV consumption it turns out that watching recorded content is by far the most popular way of nonlinear TV watching. It represents about 50 % of all nonlinear consumption in 2014.
- Time-shifted television went down by 1.6 % in 2014 with regard to the total viewing time but still it remained at an average of 6 % of the total.
- Pay TV on-demand services very much depend on the available connectivity of the broadband network used. In 2013 and 2014 on-demand pay TV viewing grew in both years by 1 %. However, pay TV on-demand only represented 13 % of nonlinear viewing and 1.7 % of total viewing in 2014.
- In 2014 OTT viewing time increased by 4.2 % across the six markets with short video content being the most requested. OTT is what made YouTube big and it is no surprise that these OTT formats are very popular among younger audiences.
- Despite the claims from online prophets around the world, online long form content constituted only 2 % of the total viewing time (or 13 % of nonlinear viewing).

Even though the study certainly has been carried out very seriously any conclusions derived therefrom should be taken with a pinch of salt. First of all, the study represents a snapshot in time and it refers to the past. It is certainly conceivable that drastic changes may happen. The changes induced with the introduction of the first generation of iPhones may be used as an example for such unpredictable rapid change. However, it is clear from the study that linear television is very strong and it may take a long time before this will significantly change.

A look at the absolute viewing times of linear TV may support this view. In Europe every individual older than 14 years was watching linear TV for about 220 min every single day in 2014 on average [EBU15a]. This impressive number of almost 4 h a day was very stable over the last years. It is true that there is a slight reduction compared to the situation some years ago but this is yet almost marginal. The situation is different when looking at the age group 14–29 where the daily consumption of linear TV is only about 150 min with a clear reduction over the last year in the order of ten percent.

Furthermore, not only the total viewing time figures confirm the relevance of linear television. The comparison between TV and other forms of media speaks for itself. According to [Zen15] linear TV has been consumed more than any other media content in Europe and the USA in the period 2010–2015 as Fig. 5.1 shows. Nevertheless, it is also true that time spent on traditional media is either stable or even declining while Internet usage is the only growing area. Detailed information on the usage of different media can also be found in [WIP15].

Figure 5.1 shows average figures including all age groups older than 14. It is fact, however, that in the group 14–29 years watching linear television is less attractive. The figures almost drop by 50 % in some countries compared to the total average. The so-called digital natives, i.e. that generation which was raised with Internet services being available whenever and wherever they needed them, show a

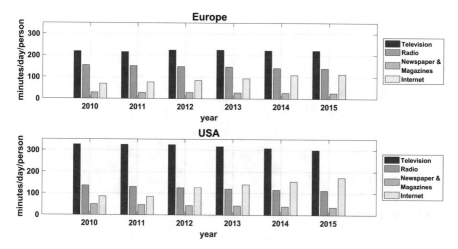

Fig. 5.1 Average daily minutes of media usage per person in Europe and the US [Zen15]

significant affinity to nonlinear and on-demand services. But still it should not be neglected that linear TV watching reaches up to 140 min in this age group which is more than 2 h a day.

New and more services are one important field of innovation for broadcasters. But also the quality of the audio-visual services produced and offered by broadcasting companies has been extended to a new level due to the availability of enhanced technical possibilities. This holds both for television and radio programmes.

The TV picture quality is determined by a set of technical parameters such as

- the spatial resolution, i.e. the number of pixels in a picture;
- the temporal resolution, i.e. the number of picture frames per second;
- the color gamut which is the range of colors that can be displayed; and
- the dynamic range representing the span between minimum and maximum brightness.

TV services were usually delivered in SDTV quality [SDT16]. In the meantime HDTV [HDT16] has become widespread even though there are still many programmes delivered in SDTV quality. Huge investments have been made to empower the whole production chain to deliver full HDTV programmes. However, the quest for higher quality has not stopped. UHDTV is a hot topic for many broadcasters in the meantime around the globe [UHD16]. It comes in two variants, a so-called 4K-UHD and 8K-UHD version. 4K-UHD is already coming to the market, there are services and TV sets available in the shops while 8K-UHD is still experimental and probably 5–10 years away.

The difference between the two UHD variants is the maximum number of pixels which can be addressed on a TV screen. SDTV has different numbers of pixels depending on the actual TV standard. However, from HD and 4K-UHD to 8K-UHD

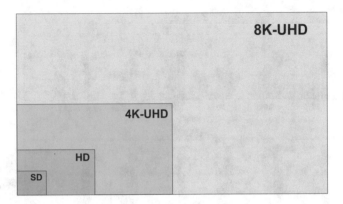

Fig. 5.2 Comparison of different spatial TV resolutions

the number of pixels grows by a factor of 4 each time. Figure 5.2 sketches the difference in spatial resolution.

However, the subjective quality perception of a viewer does not depend solely on the spatial resolution, i.e. the number of pixels. It became clear in the meantime that improving the other three parameters listed above can significantly contribute to better picture quality as well. Hence, in the discussion about how to progress with the introduction of UHDTV it turned out that fewer pixels combined with higher dynamic range and enlarged color gambit can achieve astonishing quality improvements. This is the reason why some broadcasters actually are favoring enhancing HDTV with higher dynamic range before making a huge effort to introduce UHDTV.

Improving the video quality is only one aspect of enhancing the overall user experience of television. Better sound is important too. This is definitely true as surround sound systems have become widespread. More and more speakers can be used to create a truly great watching and listening experience. But this is also very important for radio where sound constitutes the service and therefore better sound is definitely appreciated by listeners.

5.2   New Types of Devices

In order to watch a TV or radio programmes users had to use dedicated receivers for decades. At the beginning of broadcasting some hundred years ago receivers were expensive and not very handy. Hence, there was typically only one receiver per household which was located in the living room. Usually, the family met in the evenings or on Sundays and enjoyed watching TV together.

The world has completely changed since then. With the technological developments in the broadcasting sector cheaper and more capable receivers were offered on the market. This led to an increase in the number of TV and radio sets per household.

There was no longer only one in the living room. TV sets also made it into the sleeping room of the parents and in many families children got their own receivers.

Portable receivers opened the door to enjoy listening to radio almost everywhere from the kitchen and daddy's workshop to the garden. In particular very small and very cheap FM radio receivers were available which could be carried everywhere. Also radio receivers in cars became very relevant and still are important while TV reception in cars is still a niche market.

Even though the functionality of all these TV and radio receivers improved significantly they are all still dedicated broadcasting receivers. This means they can only receive particular transmissions carrying broadcast content. This content is linear and sometimes live content which users can tune in to, they may change the programme or they can switch off. In any case however, they have to watch and listen what they are offered at a given point of time.

A first level of more freedom was achieved by different types of recorders both for audio and video content. The development went from analogue tape based recording devices to digital recorders employing hard drives. Recorded radio shows or songs could be played back at any time decided by the users. Similarly, recording TV programmes allowed stripping off the restraints of linear TV programming. If at the time of the daily news show the family was not yet ready to gather in front of the TV set, the programme could just be recorded and watched later when the time was right.

The picture changed again with the proliferation of the Internet which offered to access audio-visual content also in a different way and in particular on other devices. The important issue is that these devices were not meant to be used as TV or radio receivers in the first place, yet they more and more were used for the consumption of audio-visual content as well.

Audio-visual content usage first happened on computers. Attaching a HiFi sound system to the sound card of a PC is very easy. Then, music streamed from the Internet can be listened to in a straightforward manner. Watching video content on computer screens may have been not very pleasant at the beginning but today most computer or laptop screens offer HD quality, so there is nothing to complain anymore at least from that point of view.

The game changer, however, was the introduction of smartphones and tablets. Before their appearance it was possible, to some extent at least, to listen to music on a mobile phone but it was no real pleasure. In 2007 the first generation of the iPhone was released which was a watershed in telecommunication in particular with respect to the consumption of audio-visual content. It was not about being faster than other phones, it was about a new way of providing and accessing content. Touch screens became the ultimate user interface. Instead of cumbersome operating systems and slow multi-purpose software, slim applications, the so-called apps, were put forward which represent a piece of software tailored to fulfill an exactly defined and limited purpose. The underlying fundamental principle by which all was and still is governed is user friendliness or easy usability.

Both smartphones and tablets have become so capable in the meantime with regard to computational power and display capabilities that there are no technical obstacles anymore which would prevent watching a HDTV programme on a smartphone or a tablet. The only issue left may be to get the content onto the device in a sustainable and affordable way regarding data capacity of the content delivery and the costs which come along with content access.

For many people smartphones and tablets have developed into some form of electronic interface to the world. They naturally expect that these devices can do everything they wish to do. This may be ranging from traditional voice based communication and any kind of information access to consumption of audio-visual content of various types. Eventually, this also includes consumption of all types of services broadcasters offer to their audiences. Since smartphones and tablets have become so crucial for people around the world it is clear that broadcasters need to make sure by all means that they can deliver all of their content to these devices.

There is vast number of market analyses trying to figure out how many devices are currently in use or have been sold around the planet in the past. Forecasts for the years to come are building on these data. Depending who is carrying out the analyses and forecasts and in particular on behalf of whom, the results differ significantly. However, it is clear that people are using all sorts of devices. In particular, as smartphones and tablets are gaining more and more traction this will lead to more time spent using them.

Portable and mobile IP-enabled devices go very well with the trend of a society increasingly in motion. The transformations in western civilizations with respect to work and leisure time make people become less settled. They have to move to find new work or commute between their homes and work over long distances. Clearly, this gives room to use smartphones and tablets to access audio-visual content.

While smartphones naturally are equipped with SIM cards to connect to a mobile network, this is usually not the case for tablets. Roughly more than 90 % of tablets are sold without a SIM card. Connecting to broadband network thus requires finding a usable WiFi network. When it comes to the decision how to distribute broadcast content this is an important element to be taken into account (see Chap. 8).

5.3   Usage Environments

Due to the fact that in the past the available broadcasting technology could offer only limited capabilities consumption of broadcast content took place at home, mainly in the living room where the whole family gathered when there was something interesting on air. The radio and TV receivers were bulky devices which could not easily be moved from one place to another. Furthermore, in order to enjoy broadcast content it was necessary to install a roof top antenna from which the signal was led to a fixed location inside the home. This was typically the living room.

No doubt, broadcasting receivers became smaller and more powerful over the years. Hence, they could be carried more easily and therefore gave rise to portable or even true mobile reception of broadcast content. The most prominent example certainly is the FM receiver which probably every car sold somewhere in the world is still equipped with. They work amazingly well. Only now new digital radio technologies such as DAB and DAB+ are gaining ground in Europe and Asia which may eventually replace traditional FM transmissions in the future.

Portability and mobility was always relevant for sound broadcasting. However, the situation is slightly different for television. Portable TV reception is probably quite common in the sense that nowadays there are handy TV receivers which can be moved more or less easily from one place to another to watch television. This can be, for example, also in the garden in summertime when people are having a barbecue and at the same time there is an interesting football match broadcast live by one of the TV stations.

Again the availability of smartphones and tablets in recent years has opened new opportunities to extend the individual media consumption to basically all environments and circumstances. In the past broadcasters were only concerned about offering their services for fixed reception at home. Meanwhile, all types of reception have become crucial because people want to have access to broadcast content wherever they are and whatever the environment may be. This means consumption of broadcast content takes place

- at home as it was always the case, but now very likely in all rooms from living room to kitchen, bathroom, children's rooms, workshop, garden, and even the garage;
- away from home inside of buildings in a more or less stationary situation, for example in the office or meeting room or while waiting at an airport or a train station, sitting in a restaurant or any other public building;
- away from home in the open, both in a stationary or nomadic way, for example in a stadium, a public swimming pool, or a park; and
- while being in motion either in public transport such as train, metro or bus, or in a car either driving or just sitting there with someone else driving.

From the point of view of delivering content these different environments turn out to pose very different conditions. Actually, apart from having access to different user devices, providing service under all these conditions is a real challenge both technically and economically.

An important issue to be considered with regard to usage environments is the level of control that users may have about ways to access audio-visual content. In their own flat or houses it is basically their own decision which type of broadcasting network they are making use of. They can set up a roof-top antenna for reception of terrestrial signals or they may favor putting up a satellite dish. The same applies to the decision about access to the Internet. The choice may be restricted due to factors beyond their control but the decision to go for DSL or cable broadband lies with people themselves.

This is very different in public places. There users have to employ whatever type of access to audio-visual content may be available. Away from home the only way to receive linear radio or TV programmes may be by means of terrestrial broadcasting networks. In order to access the Internet users may have to rely on free WiFi hotspots whose performance they cannot control. This has direct impact on the level of quality of service which can be enjoyed.

5.4   New Trends for the Broadcasting Sector

New types of services are coming hand in hand with increasing capabilities of networks and user devices. More capacity allows delivering more data which can be used to enhance existing services. This is in particular the case for broadband networks which constitute the basis for Internet services.

Currently there are two big trends in the broadcasting business with regard to new services. These are recommendations and personalization of services. For both interactivity seems to be indispensible as they both rest on analyzing user behavior. Only if content providers know what users are doing they can cast this into recommendations or provide the technical means to adjust the content offer to meet individual preferences.

Everyone probably has experiences with ads placed on websites which mirror the interests of the users. After having accessed an online shop searching for a new pair of shoes of a particular brand quite likely the next visit to the online shop is accompanied by ads for the latest variant of shoes of the same brand or similar ones. Having booked a room for a holiday trip in some exciting place on a hotel booking portal will usually lead to messages informing the user that there are new and now extremely cheap offers in the same location.

This is just a result of tracking the behavior and the habits of users when accessing the Internet. A huge amount of data is thus collected. This data is used to personalize the appearance of a website. Depending on their preferences two users accessing the same online portal will see different ads. Clearly, the idea is that what was interesting to users once may be interesting a second time or even permanently.

Combining the usage patterns of users allows creating recommendations. Again online shops have gained a certain level of perfection in this regard. Users are usually informed that customers who were interested in a particular object were also looking at another one or even have bought something else. This can be something completely different or also the same product but only from another company. Both types of hints are offered to users hoping that thereby sales figures can be increased.

One may ask the question what all this has to do with broadcasting. Broadcasters are also offering a product which is their radio and TV programmes. Naturally, they do not want to produce and send content which no one is interested in. Therefore, broadcasting companies always had tried to find out what their listeners and viewers would actually like to listen to or watch. In the past this information has been gathered by means of surveys or polls. Still today the broadcasting sector is heavily relying on these measures to get insight into the interests of their customers.

But clearly, emerging technologies to better understand the demands of users are very attractive for broadcasters as well. The ability to monitor what a user is watching over the span of an evening is precious information based on which programming departments of the broadcasting companies could develop more attractive content. This could even lead to specially tailored online offers for users.

However, this is exactly where the problem may start as there is an issue in particular for public service broadcasting. Linear broadcast content can be delivered by means of broadcasting technology, i.e. a one-to-all distribution mechanism which is agnostic of whether someone is receiving the content or not. Hence, reception is anonymous from the perspective of the user. No one will know if someone is watching TV or which programme is actually watched. This has been considered as one of the pillars for public service broadcasting as it allows unconstrained access to information without the possibility being tracked or even blocked. Clearly, if censorship is in place before information is transmitted there is no such thing as free access to information.

With the Internet all this is changing. First of all, now there is a possibility to collect information about user behavior in way not possible before. It is only natural that also public service broadcasters would like to exploit these new possibilities. However, it seems to be clear that there have to be certain limits which respect the privacy of users. Consequently, for public service broadcasters completely unbarred access to individual data probably seems to be a red line not to be crossed.

Another way to learn more about user habits and expectations is making intensive use of social media. This is actually done on a large scale in the meantime. No broadcaster can afford not offering traditional websites amended by, for example, Facebook or Twitter appearances. All this has become standard tools to communicate intensively with users. Social media allow getting immediate feedback to the radio or TV programme in a live manner. Clearly, also feedback referring to content which has already been broadcast is very valuable information for programming departments and programme managers.

There are many different programme formats which would not be possible without the feedback functionality of social media. Radio shows which build on getting in contact with people from all over the world are quite popular. They offer a great opportunity for people living abroad to stay in contact with their home region and get first hand information.

Social media allow creating communities which may constitute a very strong bond between the broadcasting company and its users. The element of identification with "my" station which can be brought to life through easy communication channels in both directions, i.e. from broadcaster to user and vice versa, is essential to staying successful as broadcaster.

Internet portals such as YouTube are certainly among the most successful platforms. They offer a forum to make available user generated content which has become really a huge thing. Such portals put forth a new type of media celebrities. Mostly youngsters set up their individual channels and the most successful ones attracted many million followers of the same age. The topics are usually taken from their own daily lives and presented in a way, i.e. style and language, which echoes the lifestyle of their followers.

It is interesting to note that obviously this way of presenting audio-visual content is very much appreciated by the younger generation. It is about their topics, their issues, their lives and is offered by people who could easily be their friends if they would have ever met. This creates a level of credibility which traditional broadcasters struggle to achieve. No wonder they started to analyze in detail the mechanisms for success of user generated content with the aim to adapt them for their own content offers. This reaches from developing programme formats targeting explicitly the so-called digital natives to setting up their own YouTube channels. This way they are present on these platforms thereby hoping that their adapted content will eventually reach younger audiences.

Chapter 6
Technology Developments

The last decade has seen a boost in technological development which has changed the distribution and consumption of audio-visual media services including broadcast services fundamentally. In 2007 the first iPhones hit the global markets. This was a watershed, almost a phase transition. Portable and mobile audio-visual media consumption had been targeted before already with the introduction of distribution technologies such as DVB-H [DVH16], MediaFLO [FLO16], or DMB [DMB16]. However, all of them failed. Some people argued that the primary reason for failure was the lack of promising and sustainable business models. However, based on the iPhone experience it seems more likely that the deficits of the then available portable devices were the real reason why mobile TV did not become an economic success.

Today the situation is different. Portable devices are becoming more and more powerful. They offer HD displays or better, can store a lot of data and there is basically an app for every conceivable area of life. Our way of communicating has changed entirely with all the social media allowing sharing experiences around the globe in an instant, as long as the underlying networks are capable enough supporting all communication requests. This is actually now becoming the real challenge. People are consuming more media content than ever and they are sharing all this content in various forms. Furthermore, more and more industrial sectors are being digitized thereby exponentiating the traffic on electronic communication networks.

This has triggered developments with respect to new network technologies, both fixed and mobile, which will certainly impact the way in which broadcast content will be distributed in the future. Both the broadcasting and the broadband industry are engaging in expanding the capabilities of their respective technologies in order to accommodate options to satisfy the demands of broadcasters and users.

Broadcast content is distributed across terrestrial, satellite, or cable broadcasting networks. The satellite sector seems to be very stable with respect to broadcast content distribution both in terms of technologies and business models. There are innovations which broadcasters can benefit from such as spot beams [SpB16].

© Springer International Publishing Switzerland 2017
R. Beutler, *Evolution of Broadcast Content Distribution*,
DOI 10.1007/978-3-319-45973-8_6

They allow shaping the satellite footprint to adapt it, for example, to the size of a country or to enable even regional coverage within a country. This is becoming increasingly important with respect to the need of broadcasters to meet content right obligations. However, the spot beam technology as such is not linked to the distribution of broadcast content but can be utilized whenever footprint shaping is relevant. Thus, from a broadcast technology development point of view the most interesting developments are taking place in the field of terrestrial broadcasting and broadband networks, both fixed and mobile.

6.1   Terrestrial Broadcasting Technologies

Broadcasting technologies have been the only available method to deliver broadcast content for a long time. By definition distribution on broadcasting networks corresponds to a one-to-all scheme. Linear content is transmitted from one source to an in principle unlimited number of concurrent users. Since two decades the transition from analogue to digital transmission technologies has offered new opportunities in terms of capacity and technical service quality of distributed radio or TV channels.

In contrast to broadband networks, however, the development in the broadcasting sector has taken place on a regional rather a global scale. This led to the situation that nowadays there are several digital broadcasting standards available, in particular for terrestrial television broadcasting, which are used in different regions of the world. Europe and parts of Asia employ DVB-T or DVB-T2 while in Japan and South America terrestrial television broadcasting is based on the ISDB-T standard. The USA, Canada, and Mexico rely on their own technology which is called ATSC. A detailed overview can be found at [DVB16a] (see also Sect. 3.1.1).

All DTT standards are promoted by corresponding organizations which try to further develop their technology and extend their global market share. In recent years integration of broadband features which enable on-demand capabilities has become more and more important.

Even though there is significant development taking place in the terrestrial broadcasting sector one of the biggest challenges terrestrial broadcasting technologies are facing today is the fact there is no single global standard. Globally harmonized standards open the door to economies of scale and thus reduced prices for customers. In contrast, different standards give rise to fragmentation of the markets. This is crucial with regard to the competition broadcasting technologies are facing by broadband network technologies. These are based on global standards such as the 3GPP standards for mobile broadband communication (see Sect. 6.2).

6.1.1 DVB-Project

The development and introduction of the various variants of DVB standards has been supported and guided by the DVB Project [DVB16]. The DVB Project is an alliance of about 200 companies including broadcasters, broadcasting network operators, broadcasting network infrastructure vendors, and manufacturers of TV receiver equipment. The DVB project was founded in 1993 and started out as a European project but is now active worldwide. It is hosted by the European Broadcasting Union (EBU) in Geneva.

The objective of the DVB Project is to develop and agree specifications for audio-visual media delivery based on Member initiative. This includes all types of broadcasting networks, i.e. terrestrial, satellite and cable. There is a close link with the European standards bodies CENELEC [CEN16] and ETSI [ETS16] through the so-called ETSI/EBU/CENELEC JTC Broadcast (JTC) [JTC16]. Specifications proposed and agreed by DVB Members are sent to JTC for approval. Depending on the scope the specifications are then formally standardized by either CENELEC or, in the majority of cases, ETSI.

The DVB Project rests on two pillars which are the Commercial and the Technical Module. Specification typically starts with the Commercial Module defining new features which seem to be promising from a commercial point of view without looking into any technical feasibility. This is left to the Technical Module which has to come up with proposals for technical specifications aiming to fulfill the commercial requirements.

From the very start the DVB Project was aiming at specifications which are flexible enough to accommodate new services, for example, better picture quality television formats without having the need to start all over again whenever there are new developments emerging on the content production side. Therefore, instead of developing a system which provides transmission capacity on a programme by programme basis the concept of multiplexes had been introduced at an early stage. A multiplex acts a transport container which in principle can carry SD, HD, or even UHD programmes including surround sound. On a transport layer the DVB Project has used and continues to draw extensively on standards from the ISO/IEC JTC MPEG. The transport mechanism for all systems is the MPEG2 Transport Stream.

The DVB Project has developed standards for terrestrial, satellite, and cable distribution. These are called DVB-T, DVB-S, and DVB-C. In the meantime, the second generation of these technologies, i.e. DVB-T2, DVB-S2 and DVB-C2, has been put forward and is widely used now. However, all these technologies are meant to carry linear TV content to fixed receiving installations. This is strictly true for cable and satellite while in the case of DVB-T/T2 transmission modes which allow for portable and mobile reception can also be deployed. As it was anticipated that portable and mobile reception would prove to be very relevant in the future a special terrestrial DVB variant called DVB-H ("Handheld") was developed in order to better cope with the particular challenges of portable and mobile reception. DVB-H was further developed in terms of DVB-NGH ("Next Generation Handheld"). Starting from DVB-T2 a system variant for portable and mobile reception called DVB-T2-Lite was also put forward.

Confronted with the changing habits and requirements of users the DVB-Project had to address the question how to make DVB fit for the future, in particular the terrestrial part of it. Therefore, they kicked off a study mission for the Commercial Module on the long-term future of terrestrial broadcasting. In November 2015 they delivered their report which contains the findings and conclusions of the group [DVB15]. It is an attempt to look into the future of terrestrial broadcasting and try to find trends based on which targets for future development can be identified.

Several key issues are elaborated which are considered crucial for the future success of terrestrial broadcasting. Firstly, terrestrial broadcasting is focused on the distribution of linear broadcast content. It is expected that linear TV remains strong also in the coming decade and hence DTT has its role to play there. The Commercial Module even goes further by saying that the position of terrestrial broadcasting will be consolidated due to increasing demand for high quality linear TV services such as an increasing number of HD and UHD programmes and an overall increase of the number of TV channels.

However, it is also recognized that there is a need for a global terrestrial broadcasting standard in order to keep terrestrial broadcasting competitive with respect to other forms of broadcast content distribution. Furthermore, establishing a return-channel capability is vital in order to enable access to nonlinear programme offers. The essential conclusion of the report is that DTT has to better integrate with other technologies and platforms such as mobile broadband. Otherwise terrestrial broadcasting will get locked-up in isolation and certainly lose relevance.

One of the integration issues which will certainly gain more relevance in the future is related to the question of the most future-proof transport stream mechanism. DVB so far utilizes the MPEG transport stream to carry the broadcast content. However, the world is sliding more and more towards an IP-only environment. Hence, the question arises whether DVB has to embrace IP-delivery. Therefore, the DVB project has initiated an activity which is to elaborate on possibilities going beyond transport stream in the future potentially abandoning it for the sake of going all-IP. By the end of 2016 the DVB Project should have gained a clearer view.

6.1.2   ATSC 3.0

In the USA, Canada, Mexico, and some other South American countries terrestrial television is based on the standard developed by the Advanced Television Systems Committee (ATSC) [ATS16]. ATSC is an international, non-profit organization developing standards for digital television. It has been founded in 1982. The list of member organizations of ATSC includes broadcasters, broadcast equipment manufacturers, satellite and cable companies, and various representatives of the consumer electronics, computer and semiconductor industries.

Similar to the DVB Project also in USA the broadcasting industry has been working on further developing the existing TV standard to cope with newly emerging user habits and expectations. The days when television consumption at home corresponded just to sitting in front of the TV screen to watch what broadcasting companies have decided to schedule seem to be gone. The user requirement to become more autonomous in terms of timing, location, and device could no longer be accommodated within the existing ATSC standards.

Therefore, ATSC took the decision to develop a more advanced TV standard. The target was to provide higher picture resolution (e.g., UHD), mobile reception and incorporation of broadband access. As a consequence, some basic techniques of ATSC had to be adapted, for example a renunciation of the single carrier concept by embracing multi-carrier OFDM wave forms was agreed. These have been the core of the DVB-T or ISDB-T systems from the very beginning. Indeed, on a physical layer ATSC 3.0 has incorporated many features which are implemented in DVB-T2 already.

The fundamental change, however, which discriminates ATSC 3.0 in particular from DVB-T2, is the fact that the MPEG transport stream which was part of former ATSC variants has been replaced by IP-transport in order to apply the IP protocol end-to-end. This is a drastic change entailing specifically no backward compatibility.

This decision has been justified by the claim that broadcasting would no longer be an independent silo surrounded by the Internet developments. It is expected that broadcasting and broadband delivery mechanisms could be more seamlessly connected thereby giving rise to new types of hybrid services and better catering for niche content (see, for example, [Ric15]).

Going all-IP for a broadcasting technology is indeed a controversial issue. From the viewpoint of better integration of nonlinear services into the broadcast content portfolio this may indeed by a good option. Nonlinear services rest on bidirectional communication, i.e. the possibility for a user to actively request a service. In broadband networks such communication is enabled by using the IP protocol. To this end, the data stream is split into different IP packages which are sent independently from source to receiver. If one package is missing due to bad transmission conditions, it can be requested again.

However, for linear broadcast content this is not possible and probably not even necessary. Coding and modulation of terrestrial broadcasting systems can be tuned in a way to enable optimal receiving conditions. As there is no return channel it is not possible to receive any feedback from the receiver about missing packages or bad transmission conditions which in broadband networks can be used to adapt the signal transmission.

An all-IP solution thus only makes sense if the whole chain from production to reception is IP based. If linear broadcast content production still targets MPEG transport stream, then using an IP based distribution technology such as ATSC 3.0 requires encapsulating the transport stream in IP packages. This increases the complexity without necessary adding any real benefit.

6.1.3   FOBTV

The primary objective of both the DVB Project and ATSC is to further develop and enhance their respective TV standard to make it fit for future requirements. It is true that both claim that the development of a global standard is crucial in order to remain competitive as technology in relation to broadband technologies. However, it is also true that both think their own technology should be the basis for such a globally harmonized broadcasting standard.

In 2011 another non-profit organization has been set up by 13 leading television broadcasting companies to specifically address the issue of a new global television standard. It is called "Future of Broadcast Television" (FOBTV) [FOB16]. Its members represent broadcasters, manufacturers, network operators, standardization organizations, research institutes, and others in more than 20 countries all over the world. According to its terms of reference the FOBTV initiative is aiming at developing technologies for next-generation terrestrial broadcasting systems and making recommendations to standardization organizations around the world.

Even though starting off with a lot of enthusiasm it seems that after about five years a lot of momentum has been lost. There may have been certainly many discussions in the technical groups of FOBTV but so far no visible outcome of the FOBTV activity can be noted. It remains to be seen if this is just the calm before the storm or if FOBTV is dying a silent death as many ambitious projects before.

In any case, the idea to strive for a global broadcasting standard is worth being pursued further. It may even be a matter of do or die in terms of keeping alive in particular the terrestrial broadcasting platform as competitive distribution option.

6.2   Mobile Broadband Technologies

At the beginning of the development of mobile communication systems the focus laid on voice communication. In order to call someone from home one had to use a phone which was connected to the telephone socket by cable. Outside home finding a phone booth was necessary. With the advent of mobile telephony people thus gained a lot of freedom.

New types of services based on access to the Internet could be offered only with the introduction of the third generation of mobile systems. Since then telephony seems to become less and less important for mobile phone users. Fast access to the Internet leveraged the concept of an app, i.e. slim and easy to use software applications, perfectly adapted to particular usages. In the meantime, there are apps for every conceivable purpose and it seems there is no end to it.

It was no surprise that with the increasing computational and display power of evolving portable devices the consumption of audio-visual content, including broadcast content, was reaching unprecedented levels. Actually, by far the major part of traffic on mobile networks today is due to any kind of audio-visual media consumption. Even though there is an increase of video uploading due to people sharing their own footage on social media, a very pronounced asymmetry between the amount of down- and uplink traffic can be detected.

Apart from all potential drawbacks or traffic bottlenecks which still may be experienced today, consumption of broadcast content on mobile devices such as smartphones and tablets is becoming more and more crucial, also because the relevance of these devices in the daily lives of users is still on the rise. Therefore, broadcasters started to engage in mobile broadband issues in order to inject their requirements into the development process.

6.2.1 eMBMS

Broadcasting companies are producing and offering a whole universe of different content types to their users. Even though nonlinear forms of consumptions are on the rise the most important format is still linear television and radio. In order to receive linear content dedicated receivers for fixed, portable, and even mobile receptions are available.

However, smartphones and tablets are gaining more and more importance as they become more powerful both in terms of functionality and displaying capabilities. For the users they are developing into some kind of personal electronic interface to the world. Therefore, users have the expectation that smartphones and tablets can virtually fulfill any need for communication or electronic services, including providing access to broadcast content.

Broadcasters have recognized quite some time ago the strategic relevance of providing all their service on smartphones and tablets. Depending on the traffic load of mobile broadband networks and the subscription details, user can enjoy broadcast content on portable IP-enabled devices. At least this is what the mobile network operators are promising boastfully. However, reality usually looks different. Typical subscription data caps do not allow for more than few hours of TV watching per month. Beyond this the data rate is either throttled to a level which is too small for decent TV reception or customers have to pay more to maintain sufficient data rate.

The strategy broadcasters were pursuing for a long time was to convince device manufacturers to incorporate a DTT receiver in every smartphone and tablet. There have been dongle based solutions to enable DTT reception in portable devices for quite some time. However, they never really took off commercially as it probably is just too clumsy for most users to attach some dongle in order to be able to watch TV. From a technical point of view incorporation of a DTT receiver could be done in a more or less straightforward manner as there are chipsets on the market which would allow reception of all terrestrial radio and TV standards (see, for example, [Sia16]).

This attempt failed in the past due to the fact that the market of portable devices was controlled by the mobile network operators (MNO). Subscriptions and corresponding data plans came along with subsidized phones. This way the

MNOs decided which features were integrated and which not. In the meantime, the markets are changing. In some countries such as Germany the number of unlocked phones, i.e. phones sold without any subscription, has exceeded 40 % or more. This gives manufacturers more freedom to decide themselves how to equip their devices. Unfortunately, there is still no sign that incorporation of DTT receivers is perceived as a promising add-on or a competitive edge over other manufacturers.

Another way to reach smartphones and tablets with linear content could be by means of a feature of mobile broadband networks which had been introduced already in 2008. It is called "enhanced Multimedia Broadcast Multicast Service" (eMBMS) and offers to use part of an LTE carrier as a broadcasting signal. This can be received by all mobile terminals basically in the same way as broadcasting receivers can receive a DTT signal. Actually, eMBMS is also based on an OFDM waveform. Even though standardized quite early there was very limited commercial interest to rollout such a technology by any MNO so far. Some mobile network operators in Australia, Europe, and the USA have announced plans for commercial deployments in the near future.

The possibility to make use of eMBMS for the delivery of broadcast content to smartphones and tablets triggered the European Broadcasting Union (EBU) representing European public service broadcasters to start a new initiative. The years after the World Radiocommunication Conference in 2007, i.e. WRC-07, have seen terrestrial broadcasting getting more and more under pressure by the mobile industry seeking to get hold of spectrum so far allocated to the broadcasting service. The confrontation during the battle for the 800 MHz and later the 700 MHz spectrum did not leave any room for constructive discussions between broadcasters and the mobile community about potential usage of mobile broadband networks for the delivery of broadcast content.

In order to overcome the dead-lock the EBU created a project team in 2011 to which representatives of the mobile industry were invited. The task of the new group which was called CTN-Mobile[1] was to jointly address the question "What can LTE do for broadcasting?" Several infrastructure and device manufacturers of the mobile sector followed the invitation. Unfortunately, up to now no MNO was willing to join the group.

Despite the absence of MNOs CTN-Mobile developed into a very important and constructive forum in which broadcasters and representatives of the mobile industry could meet to exchange their views and know-how without having to stick to their respective public roles both have to play when attending meetings of the ITU or CEPT. It has to be emphasized though that the key to successful creation of a faithful environment was the decision not to talk about spectrum, at least not spectrum allocations. Rather, the focus of the work should lie on technical and business related aspects trying to elaborate under what conditions LTE could help to fulfill the objectives of broadcasting companies.

---

[1] CTN stands for *C*ooperative *T*errestrial *N*etworks and corresponds to the name of the mother group of EBU Members to which this project team was attached to. Mobile is to indicate that the scope is mobile broadband networks.

CTN-Mobile published a first report in 2014 which deals with several technical aspects of broadcast content delivery over LTE networks [EBU14a]. The study revealed that some of the basic requirements of broadcasters could be met already today under certain conditions. This refers, for example, to the provision of linear free-to-air services. Every LTE user terminal establishes access to a network through a SIM card which is linked to a corresponding subscription to an MNO. Reception of FTA television would require specifically configuring the SIM card by the MNO. The associated regulatory, operational, and business aspects of this have not been addressed in the report.

The performance of an LTE eMBMS system was analyzed based on various studies. A number of issues have a significant impact on the performance, i.e. influence the achievable spectral efficiency. These are primarily related to network design and coverage targets. Even though spectrum allocation issues have been excluded from the scope of CTN-Mobile there have been extensive discussions about spectral efficiency as this will directly determine the amount of spectrum which would be necessary to implement such kind of delivery of broadcast content.

In the course of the discussion it became evident that broadcasters and mobile industry are employing different methodologies to assess the performance of a terrestrial network. Therefore, it was necessary to develop a common understanding how the network performance with regard to achieving the coverage targets should be assessed. CTN-Mobile published a second report listing and describing in detail all elements necessary for the simulation of LTE networks aiming to offer broadcast content [EBU15b].

Very controversial discussions took place in CTN-Mobile on the costs for delivery of linear broadcast content over LTE eMBMS networks. Both broadcasters and the mobile industry presented their first estimates which actually diverged by almost an order of magnitude. No conclusion could be reached for which reason the CTN-Mobile reports do not contain any conclusive statements about expected distribution costs. Actually, this was one of the topics where the participation of mobile network operators could have been beneficial in order to get first hand input.

CTN-Mobile not only proved to be a very important forum to jointly discuss LTE issues. Activities in other groups or organizations such as ITU or 3GPP (see Sect. 6.2.2) were also discussed. In the preparation for other meetings CTN-Mobile turned into a very helpful platform to align positions and draft input contributions in those cases where there was common ground between broadcasters and the mobile community. This also refers to eMBMS trials which have been carried out in recent years.

Trials on eMBMS have been carried out, for example, by Verizon on the occasion of the Super Bowl in New York [Ver14]. The objective of this test was to provide different types of video content to the visitors in the stadium as an additional service to the live experience. Views from different camera angles or slow motion sequences of particular scenes of the match could be made available to the users within the stadium.

In Paris, France and in the Aosta Valley in Italy a special approach developed by the University of Braunschweig has been under investigation by TDF and RAI

[TDF15]. The so-called LTE-A+ technology makes use of the future extension frame feature of DVB-T2 which allows embedding (modified) eMBMS data. These data can be directly received and displayed by correspondingly enabled LTE user devices.

Probably the most advanced eMBMS trial dealing with delivery of broadcast content over LTE networks was carried out in Munich, Germany between 2014 and 2016 under the leadership of the Institut für Rundfunktechnik (IRT) [IRT16]. IRT is the R&D institute of the German public service broadcasters. The project called IMB5 ("Integration of mobile and broadcast radio in LTE/5G") focused on investigating the capabilities of eMBMS operated in a single frequency network using the state-of-the-art eMBMS technology. Existing terrestrial broadcasting stations were used to build the transmitter network.

The aim of the IMB5 project was to explore the possibilities and limits of the currently standardized eMBMS service of LTE. From a broadcasting point of view ubiquitous coverage with a guaranteed quality of service is crucial. The trial was meant to give an answer to what extent existing eMBMS technology could offer this and where extensions are needed. In order to achieve in particular large area coverage for linear broadcast content a large cell network infrastructure, i.e. with an individual cell radius larger than 15km, seems to be beneficial in terms of efficiency and costs.

However, as broadcasters are also interested in offering nonlinear services which build on a unicast return-channel capability, a combination of large and small cells should be envisaged. The Munich trial confirmed that eMBMS may become a promising way to deliver both linear and nonlinear content to smartphones and tablets. But it seems the current specifications still call for enhancements to fully comply with the requirements of broadcasters.

6.2.2   3GPP TV Enhancements for LTE

The specification of further developments of new generations of mobile networks is carried out under the auspices of the 3rd Generation Partnership Project (3GPP) [3GP16]. It provides a forum to different telecommunications standard development organizations around the world to produce the Reports and Specifications that define 3GPP technologies.

The project works to further develop cellular telecommunications network technologies. The main working areas are radio access technology, the core transport network, and service capabilities which are reflected in the working group structure of 3GPP. Specifications and studies are contribution-driven in Working Groups and at the Technical Specification Group level. Clearly, work on codecs, security and quality of service are integral part of the specification effort. Hence, 3GPP specifications offer an end-to-end system all under control of 3GPP. The specifications also provide interfaces for non-radio access to the core network, and for interworking with WiFi networks.

The end-to-end character of 3GPP proved to be a very successful approach with respect to mobile communication of individuals. Offering phone calls, text messaging, and access to the Internet have been the basis on which mobile networks spread out all over the world. However, most MNOs have realized in the meantime that providing connectivity alone is no longer a future-proof business model. This is the reason why more and more MNOs seek to provide services beyond connectivity. Access to audio-visual premium content is one of the promising areas. This is in particular relevant as the last years have seen an enormous increase in data traffic on mobile network caused by consumption of video content. Therefore, it is self-evident to try to gain a foothold in the market of audio-visual content provision.

One of the major players in the audio-visual content sector is broadcasting. Therefore, it was quite logical that 3GPP started an initiative in early 2015 to enhance the eMBMS functionality of LTE in order to enable any kind of TV services on 3GPP systems. When EBU took note of this activity it was decided that it is strategically important to engage in this process in the light of making available all broadcast content, i.e. linear and nonlinear, on smartphone and tablets. As all the attempts to incorporate DTT receivers in mobile terminals in the past had more or less failed, enhancing eMBMS to facilitate TV services could open an even more efficient possibility of getting access to portable and mobile IP-enabled devices.

In order to facilitate new specifications that would accommodate the requirements of public service broadcasters, EBU came to the conclusion that it is crucial to actively participate in this specification activity from the very beginning. Consequently, EBU supported by some manufacturers and even MNOs contributed by requesting that the future 3GPP system shall

- allow for free-to-air delivery of broadcast content;
- envisage a receive-only device, i.e. no authorization at the network;
- overcome the limitation that only up to 60 % of a carrier can be used for eMBMS;
- provide the mechanism to deploy a standalone eMBMS network without any unicast elements;
- facilitate mixed eMBMS and unicast usage on different carriers or even networks;
- offer the possibility to provide a guaranteed quality of service across the entire envisaged coverage area; and
- render possible establishment of large coverage areas, ranging from regional to national.

These proposals derive from the basic requirements of public service broadcasters which any relevant distribution technology has to meet in order to fulfill their remit (see Chap. 7). Free-to-air distribution is not debatable for PSBs as it is in most cases a binding legal obligation. Removal of the 60 % limit is trivial from a PSB point of view in order to utilize resources in a most efficient manner.

Efficient usage of resources is also the basis of the proposal to enable standalone eMBMS networks. In most countries there is more than one MNO. If the broadcasting community would wish to use eMBMS to deliver tens of TV channels, for example in HD quality, this would require significant spectrum and network resources. In order to reach the whole population each MNO would have to provide all of this content to its own subscribers which represent only a part of the population. Thus, the same content would be distributed several times. This is without any doubt a waste of resources let alone leading to prohibitive distribution costs for broadcasters.

Therefore, a dedicated standalone eMBMS network could be established which is exclusively used to deliver the whole set of linear TV programmes. This network could then be shared by all MNOs through allowing all their subscribers access to it. Who is operating such a network becomes a secondary question. It could be a traditional MNO but in principle it could also be a broadcasting network operator who thereby could extend his portfolio to offer content delivery to smartphones and tablets.

If a broadcasting network operator would be running a standalone eMBMS network, they would certainly try to build on existing broadcasting network infrastructure. Terrestrial broadcasting networks are providing services across large areas. In the case of DTT networks this is typically achieved by using a limited number of so-called high-power-high-tower stations (HPHT). HPHT transmitters are operating at significant effective radiated output power which can reach 100 kW or more. This gives rise to a large coverage area per transmitter, or in mobile technology language large cells. Therefore, the distance between next-neighboring stations, the so-called inter-site distance (ISD), is correspondingly large.

In contrast, mobile networks are based on a low-power-low-tower network topology (LPLT). This results in cells with small radii. In order to cover a large area many stations with a correspondingly small ISD need to be employed.

Large area coverage by means of digital terrestrial broadcasting systems is usually accomplished in terms of single frequency networks. This means that all transmitters in the network use the same frequency range to provide the same content. The major issue of single frequency networks is to avoid self-interference. Self-interference free SFN operation using OFDM waveforms requires a guard interval, or in 3GPP language a cyclic prefix, of appropriate length which has to be adapted to the inter-site distances in the network.

The current 3GPP standard, i.e. 3GPP Release 12, provides only cyclic prefixes of up to 33μs. For a self-interference free SFN network the next-neighbor distance between two transmitters should therefore not exceed about 10km. This is not sufficient for the case of a broadcasting network operator who would like to deploy a standalone eMBMS network based on existing HPHT broadcasting sites. In this case, the next-neighbor inter-site distance can reach up to 100km. As a consequence, the cyclic prefix of 3GPP would need to be extended to a similar magnitude as for DTT in order to allow using DTT HPHT locations.

It did not come as a big surprise that some of the proposals of broadcasters into 3GPP were not very welcome by some MNOs. Amazingly, the free-to-air issue was not as contentious as it might have been expected. It was rather the no-authorization and the standalone network which triggered a lot of resistance. The motive for this seems to be pretty clear. As 3GPP networks have always been end-to-end enterprises under the control of MNOs both requests undermine the position of MNOs. Some claimed that they will never accept any economically viable traffic bypassing their SIM cards. Reception of content without being registered is obviously perceived as a direct assault on their business model.

After lengthy discussions most of broadcaster's requirements made it into normative text on service requirements. This is the first step on the way to a full specification. Clearly, this was only possible in a concerted effort by broadcasters, manufacturers and some MNOs. The target of the enhancements of eMBMS for TV services is Release 14 which is expected to be released mid-2017. It still remains to be seen if this can indeed be achieved or if a delay to Release 15 will have to be accepted. This will be out about 18 month later.

All this very much depends on available resources within 3GPP and, without any doubt, on the interest of 3GPP members to engage in the specification process. In particular, MNOs need to understand that delivery of TV services over 3GPP networks bears new business opportunities. So far, for most MNOs revenues are generated through user subscriptions. Hence, there is a business relation between the MNO and the user but not between content provider and MNO. In a free-to-air delivery context this will be reversed. MNOs have to ensure access to FTA broadcast content for users at no additional costs for them while the business arrangement now is established between broadcast content provider and MNO (see Sect. 8.5).

6.3 5G: Next Generation Mobile Networks

Mobile communication technology was introduced more than 30 years ago at the beginning of the 80s of the last century. Since then the world has seen the development and roll-out of four generations of mobile networks. While at the very beginning the quantum leap in technological development consisted in providing the ability to make a phone call while being on the move, the added value of mobile communication has moved to Internet access with all its diverse services in the meantime. Table 6.1 gives an overview about the different generations.

Now, the next generation of mobile technology is taken aim at by different stakeholders around the globe. It runs under the label "5G" and seems to be the topic which fans the wildest dreams of politicians, manufacturers, and researchers.

Table 6.1 Features of mobile technology generations

Generation	1G	2G	3G	4G
Services	Analogue phone calls	Voice, SMS, MMS	Voice, SMS, MMS, Internet access, video calls	SMS, MMS, Internet access, AV media, cloud services
Standards	AMPS, TACS	GSM, GPRS, EDGE	UMTS, HSPA	LTE, LTE Advanced
Data rates	No data	< 100kbits/s	Up to 2MBits/s	> 10 Mbits/s
Market launch	Early 80s	1991	2001	2010

In particular for politicians and regulators in Europe 5G seems to be a way to regain global leadership in the mobile technology markets. Whilst Europe was the pacemaker at the time when 2G technologies were developed and rolled-out around the world, European industry has missed the train to prosperity during the following evolution cycles which came with 3G and 4G network deployment. With 5G European politicians sense an opportunity for the European mobile industry to recuperate in particular as they anticipate fundamental change taking place in the telecommunication sector. This is reflected in the vision of the European Commission [EuC14]:

> 5G is a new network technology and infrastructure that will bring the capacities needed to cope with the massive growth in the use of communication—especially wireless—technologies by humans and by machines. 5G won't just be faster, it will bring new functionalities and applications with high social and economic value.

Apart from the expectation that there will be high societal and economic impact the most important issue is that 5G no longer refers to communication between humans only. Rather, the radically new thing is that machine type communication will dominate telecommunication in the future. Indeed, machine-to-machine (M2M) communication may be beneficial in basically every sector of modern societies which rely on computers and automation of some kind. M2M communication quite often is also addressed under the label "Internet of Things" (IoT). In any case, 5G is the first mobile technology to focus from the outset on providing services beyond voice telephony or mobile broadband access.

As the European Commission expects 5G to spread out into all economic areas they started a process to proactively engage different sectors. This refers in the first place to health, energy, transport, infrastructure, and media. These so-called "verticals" have been invited to collaborate in the definition of the basic features of 5G technology and become part of technological development process [EuC15b].

Prominent examples of how 5G is expected to influence future societies are:

- **e-health**: This includes remote surgery where the surgeon in charge is not physically present in the operating room but is steering a surgical robot through remote control. UHD video transmissions and real-time performance of the systems are vital.
- **smart cities**: The concept of smart cities targets to integrate information and communication technology solutions to manage a city's assets in a secure fashion. This includes all public institutions and community services such as schools, transportation systems, hospitals, power plants, water supply networks, waste management, and law enforcement. The goal is to improve quality of life and increase the efficiency of services. A large set of sensors deployed at relevant locations and in buildings allows collecting data which can be used to steer urban processes.
- **smart grids**: Sustainable production and delivery of energy is one of the biggest challenges of future societies. Similar to the smart city concept uniform collection of data about energy consumption can be used to steer the process of energy production and storage in order to increase efficiency and reduce harmful emission in particular if green energy components are integrated.

- **autonomous driving**: Big effort is undertaken to develop cars which are able to safely drive without any driver intervention. This requires permanent monitoring of the immediate and larger vicinity of the car. Very low latency car-to-car communication is therefore required together with access to additional information about traffic situations as well as weather and road conditions.
- **enhanced media and entertainment**: There is an increasing request for better quality audio-visual content as well as access to media everywhere, under all receiving conditions and on all available devices. Multi-player gaming has become one of the drivers of innovation in the field of media.

These are just some of the examples which are currently discussed under the 5G label. There are some elements which are common to all of them, however. In all cases the densities of devices communicating to each other either on a device-to-device basis or through a central entity such as a base station are likely to explode in number. Secondly, it can be expected that there will be a huge numbers of sensors and monitoring devices being deployed to track every aspect of the future societies. This will certainly give rise to fierce debates about data security and privacy issues which may pose real challenges to modern open-minded communities.

Sketching a vision about a bright future is one thing, casting it into real technology is something else. This is currently pursued by many organizations worldwide in parallel such as 3GPP or ITU. All big players in field of telecommunications, i.e. mobile network operators and infrastructure and device manufacturers are putting a lot of effort into technological developments. There are high-flying promises about when the first 5G networks will be deployed. It remains to be seen if this all comes true. In any case, 5G has developed such a momentum that basically no research proposal in the field of telecommunication can expect any financial funding if there is no relation to 5G activities.

Without specification and standardization of the new technologies there will be no new business opportunities. Moreover, standards have to be globally applicable in order to achieve economies of scale to offer technologies on the markets at competitive prices. Therefore, in particular manufacturers are heavily engaged in the process with the aim to get their patents become part of the future standards to safeguard their revenues.

5G shall not only provide higher data rates. There is also a need for significantly lower end-to-end latencies in order to embrace real-time applications as much as possible. Serving more than ever devices concurrently and being able to offer basically unlimited network capacity are complementing requirements. All that, however, shall be enabled with energy efficiency being reduced by orders of magnitude at the same time. Figure 6.1 sketches the basic technological requirements of 5G.

The propositions of the European Commission could give the impression that 5G will be a completely new system with no connection to existing mobile system generations whatsoever. Some manufacturers have a more differentiated view. In their White Paper on 5G Radio Access Ericsson, for example, comes to the conclusion that [Eri16]

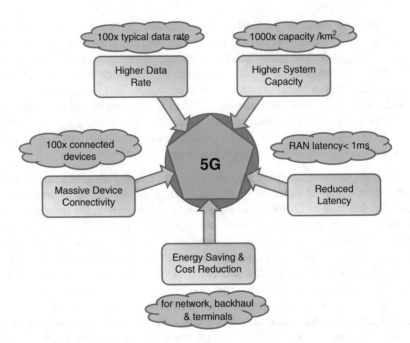

Fig. 6.1 Technical Requirements regarding different features for 5G systems

5G wireless access will be realized by the evolution of LTE for existing spectrum in combination with new radio access technologies that primarily target new spectrum. Key technology components of 5G wireless access include access/backhaul integration, device-to-device communication, flexible duplex, flexible spectrum usage, multi-antenna transmission, ultra-lean design, and user/control separation.

This view is basically supported by Huawei who also in a White paper on 5G explicate that [Hua13]

5G wireless networks will support 1,000-fold gains in capacity, connections for at least 100 billion devices, and a 10 Gb/s individual user experience capable of extremely low latency and response times. Deployment of these networks will emerge between 2020 and 2030. 5G radio access will be built upon both new radio access technologies (RAT) and evolved existing wireless technologies (LTE, HSPA, GSM and WiFi). Breakthroughs in wireless network innovation will also drive economic and societal growth in entirely new ways. 5G will realize networks capable of providing zero-distance connectivity between people and connected machines.

The common ground is the unification of different technologies under one roof. This has been carried over to the work of Working Party 5D of the ITU-R which published an ITU-R Recommendation on the development of what the ITU calls IMT-2020 [ITU15a]. IMT stands for "International Mobile Telecommunications" and is nothing but the ITU phrase for mobile systems such as the 3GPP system, i.e. 3G, 4G, or now 5G. Recommendation ITU-R M.2083 has been developed with the participation of representatives of the mobile industry and shall constitute a framework and a guideline for the development of the technical specification of the 5G system. It has also been brought to the attention of 3GPP in order to align their activities with the discussions in ITU-R.

Pushing the limits of basic technical parameters is only one part of the innovation which is envisaged in the 5G ecosystem. The idea to serve different vertical markets at the same time gives rise to organizing the network resources in terms of so-called slices. Different verticals may have very different demands. Tailoring solutions to satisfy these demands shall be based on using logical resources of a single network rather than physical resources of different networks. The network slicing concept is actually nothing really new. However, 5G shall embrace it from the very beginning in order to provide connectivity in a way that is both highly scalable and programmable—in terms of speed, capacity, security, reliability, availability, latency, and impact on battery life [Eri15].

This shall be further be enhanced by combining network slicing with software-defined networking (SDN) and network functions virtualization (NFV) which both are meant to decouple physical network components from actual tasks in order to create a more agile network. As a consequence, a single physical network is intended to respond to and serve completely different types of service requests. Very likely, all that could be in the hand of a single operator following Ericsson's 5G slogan "one network–multiple industries" [Eri15] which from a gate-keeping point of view could be considered a threat for broadcasters.

Whatever form the 5G system will finally take in terms of technical features, for example, on the level of wave forms, if all requirements are to be fulfilled then many different existing and yet to be developed technologies will need to interact. Therefore, one of the big challenges lies in the ability of different 5G implementations to seamlessly interoperate. This calls for very flexible and powerful interfaces to be developed. In particular, interfaces connecting legacy LTE technology with completely new defined 5G elements have to be very carefully specified.

There is no doubt that 5G will also have an impact on the distribution of broadcast content. If 5G is to develop into the dominant means of electronic communication, then certainly broadcasters may wish to participate. They are offering a growing variety of different services which are consumed by users on diverse devices and under various receiving conditions. In order to satisfy all these demands broadcasters are employing all existing distribution means as appropriate which is sketched in Fig. 6.2. Whether or not all these different distribution mechanisms are replaced by 5G technology in the end is the big question.

In any case, the 5G development is an activity broadcasters have to closely follow. On one side it may give rise to a new way of distributing broadcast content thereby having the opportunity to offer a greater variety of content and hopefully reduce distribution costs. Furthermore, this process is strongly supported by politics and industry, so it may well be successful. On the other hand, 5G is an area which will certainly be dominated by telecommunication companies of various kinds. New types of business models are likely to emerge. Hence, the question arises how this will impact the role of broadcasting, in particular the role of public service broadcasting.

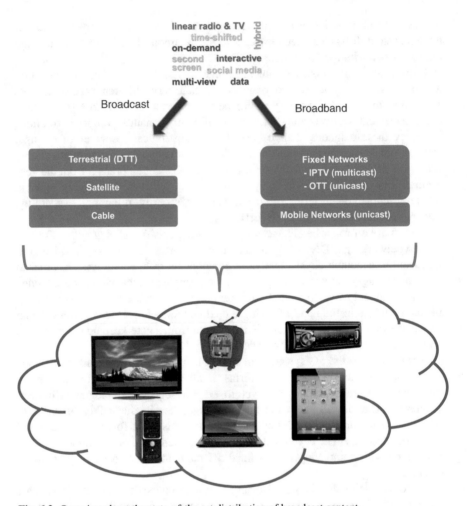

Fig. 6.2 Overview about the state-of-the-art distribution of broadcast content

At the moment it is far from being clear if the full-bodied promises wandering around will be coming true. It seems that in organizations such as 3GPP or the Working Party 5D of the ITU-R the term 5G is primarily associated with wireless communication. However, it may well be that real innovation is waiting on the wired network side. If the world will indeed see the envisaged explosion in deployment of 5G enabled telecommunication devices, for example in terms of sensors and meters almost everywhere, then it is also clear that all this traffic has to be carried over corresponding fixed backhaul and backbone networks. Therefore, it seems that 5G will primarily boost the development of fixed rather than wireless infrastructure.

As a consequence, there is a need to significantly increase the role-out of fiber networks. Clearly, all base stations need to be connected with properly dimensioned fiber links. But fiber networks also have to come closer to the location where all this data traffic is generated, i.e. at home, in offices and factories. Naturally, any kind of fixed reception or indoor communication will be routed over these fiber networks.

No doubt, wireless links will be necessary in the future in particular as portable and mobile devices will play a central role. There is no other possibility to reach a portable or mobile device than by means of a wireless link. However, these wireless links have to be considered as wireless extensions to the omnipresent fiber-based fixed network. The only remaining question thus will be how far away the next access point to the fiber network will be. Is the access around the next corner, on the other side of the street or a few blocks down the road? Such a scenario does not seem to be science fiction, at least not for urban areas. Outside the cities the situation may still be different for some time.

Clearly, such a quantum leap in infrastructure development will have a significant impact on the role mobile network operators and their business models. The question is whether existing business models of MNOs are sustainable under such conditions. This is particularly interesting when looking at the life cycles of previous mobile technology generations.

GSMA, the organization representing global mobile network operators and manufacturers, gives insight into the dynamics of the mobile developments triggering some questions. In [GSM14] an overview can be found about the time when a mobile generation was launched and at which point later in time it reached its maximum market penetration. There seems to be a remarkable pattern which says that new generations roughly are launched every ten years, while a particular generation reaches its peak penetration only about twenty years after its launch.

This means that in 2020 when 5G is to be launched 3G will just have reached its peak and 4G is still growing. Even though manufacturers will be happy about such a situation because they may sell more devices and infrastructure than ever, the MNOs may struggle. Introducing a new mobile technology entails huge investments which have to be refinanced over the life cycle of the system. This may pose a real challenge if several technologies have to be sustained in parallel over a longer period of time. Maybe this is one of the reasons why different levels of enthusiasm can be detected between manufacturers and mobile network operators.

6.4 Hybrid Broadband Networks

Mobile network operators have been very successful to provide voice and text services to users around the world for a long time. Since about a decade offering access to the Internet has been another branch of prosper business opportunities. Different mobile broadband technologies have been utilized thereto.

However, it has also become apparent in recent years that providing connectivity alone is no longer a future-proof business model for MNOs. Indeed, offering content, mainly audio-visual content, has become more and more attractive to operators. To this end, they started to make an effort to get exclusive access to premium content. Since quite some time mergers between different companies can be observed. Mobile network operators buying cable TV operators, fixed broadband and even satellite operators or vice versa. This goes hand in hand with the acquisition of Pay TV content providers.

All these company mergers target to combine wireless and wired network infrastructure together with access to premium audio-visual (pay) content in one hand. As a consequence, operators significantly increase their range and can provide an attractive content offer at the same time. From a broadcasting point of view, in particular from a public broadcasting perspective, this gives cause for concern as new and very powerful competitors may emerge fighting for the attention of customers.

This kind of combination of different network topologies may be considered as a first step on the way to truly hybrid broadband networks. However, these trends just appear to be business related workarounds in contrast to some projects which have been initiated some time ago. The latter, pursued, for example, by Google and Facebook, are likely to lift hybrid networks to a new level.

Google has been the synonym for search engines for a long time. Around that a plethora of different applications have been created which are enjoyed by users around the world. Offering cloud services and computational power through their server farms tops off the Google service range. However, in recent years Google also started to engage in development and roll-out of distribution infrastructure. There are two major projects which may have a significant impact on the future distribution network landscape. These are Google fiber [Goo16] and Google Loon [Goo16a].

Google fiber targets to roll-out fiber networks in the USA. It started in middle size cities and is now being extended to larger urban areas such as San Francisco or Los Angeles. Ultra-fast broadband connections shall be offered to customers on a fiber-to-home basis [Goo14]. This way any type of service shall be accessible inside the homes covering all kind of audio-visual Internet usages up to UHDTV.

Google Loon can be considered as some kind of complement in the sense that it targets to provide broadband access primarily in rural areas. Project Loon consists of a network of balloons traveling on the edge of space. These are complemented by solar powered drones cruising at heights of up to 20 km above ground. Balloons and drones are realizations of what is typically called "high-altitude-platform-systems" (HAPS). HAPS have been an issue in the spectrum discussions of ITU-R for years but it seems only now the technology is mature enough to be seriously deployed. According to Google the project Loon is targeting primarily those areas around the globe where people so far do not have any possibility to connect to the Internet. However, it is obvious that it will be used in the USA and Europe as well as soon as business cases can be made.

Google is not the only global player who is engaging in new broadband infrastructure. Also Facebook has started similar activities. Under the label "Internet.org" [Fac16] Facebook is pursuing an agenda which first of all is meant to provide Internet access to those people around the world which are yet not connected. Part of Internet.org is an activity called Connectivity Lab that very much resembles Google's project Loon. High flying drones powered by solar cells and equipped with very efficient electric engines are to provide access to the Internet in remote areas. Connection between airborne drones and earth stations is based on infrared laser links.

But Internet.org is not only about deploying drone based broadband infrastructure. Facebook initiated cooperation with several chip and device manufacturers to offer cheap, thus affordable devices in developing countries. Following the official statements about the project it seems there is a certain level of altruism driving this activity. However, Facebook as all the involved companies are no charity organizations. Hence, vital business interests are certainly part of the engagement in Internet.org.

The engagements of both Google and Facebook are remarkable in the sense that both companies decided to spend money on developing and rolling-out broadband infrastructure themselves rather than waiting for the traditional network operators to get going. It seems they are no longer satisfied with the slow progress or they decided it may be of strategic importance to lead the design of future network infrastructure. In the light of the influence and the economic power of both Google and Facebook dominating the world already today, their engagement may create further challenges. In particular the question of regulation of such networks may pose severe issues for public service broadcasters when it comes to finding appropriate distribution options for their content.

Chapter 7
Scope of Broadcasting

There are different types of broadcasting companies which are following different interests. They basically fall into two categories: those are commercial broadcasters and public service broadcasters. In fact, they are operating under different commercial conditions and constraints and may also be subject to different regulatory obligations.

Traditionally, commercial broadcasting is based on revenues from advertisement. There are both national and regional commercial broadcasters which target at customers wishing to advertise across a whole country or only a limited geographical area.

National commercial broadcasters are very much interested in big events which are likely to attract large audiences. This can be famous sports events, elections, Hollywood movies, or popular TV series. Depending on the demand for particular TV content the price for a few seconds of advertisement time seems to be unlimited. Examples are the UEFA Champions League, the Football World Cup, the Olympics or the US Super-Bowl. For the latter several million dollars for a 30 s commercial are standard in the meantime.

For regional or local commercial broadcasters the situation is different. They have to rely on customers which want to reach audiences in the areas where their regional business interests run. A furniture store which attracts only customers within a radius of less than 100 km does not want to spend a lot of money on ads being broadcast throughout the entire country. Therefore, quite often radio advertisement is more attractive to these regionally engaged businesses than placing commercials on regional or even local TV programmes.

In addition to advertisement based commercial broadcasting, subscription based TV and radio offers are gaining ground around the world. Most of the time, subscription to a broadcast service comes along with the absence of advertisement. Hence, those audiences which are sick of getting interrupted every 15–20 min by a set of commercials may decide to opt for a subscription of certain TV or radio channels if the price is right.

© Springer International Publishing Switzerland 2017
R. Beutler, *Evolution of Broadcast Content Distribution*,
DOI 10.1007/978-3-319-45973-8_7

The fact that advertisement or subscription based broadcasting is a proper business as any other commercial activity has certainly an impact on the content distribution strategies. A commercial broadcaster will only decide to employ a particular distribution technology if there is a clear economic reason to do so. If a distribution technology cannot reach those audiences which constitute the primary target of the transmitted commercials, then it will just not be used.

The situation is very different for public service broadcasting both from a business point of view and when it comes to the question of content distribution. Broadcasting is regulated whether it is commercial or public but the regulatory obligations of public service broadcasting are usually broader and stricter.

7.1   Public Service Broadcasting

The primary objective of public service broadcasting (PSB) is to serve the general public with broadcast content. This includes in the first place linear radio and TV programmes. However, with the proliferation of access to the Internet the portfolio of content offered by public service broadcasters has fundamentally changed. A rich variety of different audio-visual content offers such as online audio-visual content, streaming offers, and the entire set of dedicated accompanying social media appearances have become an integral part of public service broadcaster's choice. Since this goes beyond traditional programming sometimes also the term public service media (PSM) is used instead of public service broadcasting.

Through PSB citizens are informed, educated, and also entertained. Public service broadcasting shall neither be commercial nor state-owned, free from political interference and pressure from commercial forces. When pluralism, programming diversity, editorial independence, appropriate funding, accountability and transparency of PSB are ensured, it can serve as a cornerstone of democracy.

In Europe and many other parts of the world, PSB organizations are influential market players which have to coexist with commercial audio-visual service providers and compete with them for the attention of the audience. The market share of PSB is different across different regions of the world. While in Europe in some countries it can go up to more than 70 %, PSB seems to be almost marginal in the USA for example.

The status and obligations of PSB are regulated by law. Some of the key PSB obligations include providing radio and television content for all members of the society and delivering it free-of-charge. The first obligation can only be fulfilled if broadcast content and services are universally available and prominently placed so that they are easy to find and use. This task usually is captured by the term universal service obligation.

In the past when the only viable distribution technology available was terrestrial broadcasting the situation was quite clear. First of all, analogue terrestrial broadcasting only had very limited capacity. Thus, the market of radio or TV programmes was small. Hence, findability of public service broadcast content was not an issue. Furthermore, many broadcasters operated the terrestrial networks themselves. Thereby, the achievable coverage was under their control and the universal coverage obligation could be met in a straightforward manner. It was an issue of budgets for network deployment safeguarded by a corresponding regulatory framework, but it was not so much an issue of competition and gate-keeping.

Nowadays the situation is different as there are many more distribution paths. Since these forms of distribution are no longer controlled by broadcasters they may not offer in all cases what broadcasters need to have. Either they cannot facilitate the right coverage area or they are too expensive. Thus, broadcasters are forced to make use of a set of different complementary distribution mechanisms in order to reach their audiences. Quite often a combination of distribution including terrestrial, cable, satellite, and broadband networks is employed.

In most cases it is up to the broadcasting companies themselves to shape their distribution. Nevertheless, in some countries in Europe there is regulation in place which forces public service broadcasters to meet the universal coverage obligation in terms of making use of terrestrial broadcasting networks. Under these circumstances the terrestrial networks shall target to reach full area coverage or at least provide coverage to citizens wherever they live. In other countries this has been relaxed in the sense that as long as radio and television programmes of public services broadcasters can be received through at least one distribution method by everybody then universal coverage is achieved.

This relaxed interpretation of meeting the universal coverage obligation may nevertheless be in conflict with the second basic principle public service broadcasters have to adhere to which is providing content free-off-charge. This actually means that users shall not have to bear additional costs to access PSB content beyond what they have to spend on their obligatory license fee and the costs for receiving devices and infrastructure already. This is typically called free-to-air (FTA) distribution.

FTA delivery of radio and TV programmes is not linked to a particular form of distribution. In principle, FTA access to content can be enabled in all networks. However, it is usually associated with terrestrial broadcasting. This does not mean that terrestrial broadcasting is only FTA broadcasting. Indeed, in many countries access to content delivered on terrestrial networks is based on special subscriptions. However, at least in Europe terrestrial networks offer access to the content of public service broadcasters on a free-to-air basis in more or less all countries.

A pre-requisite to FTA access is that content is not encrypted. Without encryption users can enjoy broadcast content as soon as they have purchased and installed the necessary receiving equipment, i.e. appropriate antennas and receivers. Users have different options they can choose from. There are TV sets on the market with integrated receivers or set-top boxes which are attached to a TV screen in order to display content. FTA distribution of broadcast content does not require any special hard- or software to unlock the programmes, i.e. there is no need for conditional access cards or proprietary receiving devices. Also, FTA excludes any kind of subscription or pay-per-view schemes.

A variant of access to content which is closely related to FTA is called free-to-view (FTV). In contrast to FTA free-to-view means that there are no additional access costs for users similar to FTA, however, access to the content is barred on a technical level because radio and TV programmes are encrypted as they are for a subscription service. In order to unlock the access to the content users have to make use of corresponding hard- or software. This can be, for example, a conditional access card for a set-top box. In the case of FTV such a card would be made available to users for free subject to certain conditions. These could be, for example, that they are residents of the country and they can proof they pay their license fee. Whether or not FTV is a viable way to grant access to PSB content depends on the national regulation. In Germany, for example, public service broadcasters can make use of FTV distribution as long as there is still an FTA alternative available.

It is important to highlight that free-to-air does not mean there are no cost for users. Countries which offer PSB services impose either a special tax on their citizens for this or there is a license fee to be paid and collected by an independent entity. There are different approaches in place how this license fee model is implemented. However, one element is common to all of them. Even though in the perception of the public payment of the license fee is associated to getting access to PSB content, i.e. for the delivery of radio and TV programmes to people, most of the budget generated by the license fee is actually fed into production of content.

In the analogue days when the media world was clearly structured it was necessary to use a dedicated broadcast receiver to listen to radio or watch a television programme. Hence, it was straightforward to develop a model for the license fee in which those people who possessed a radio or TV receiver had to pay a certain amount of money every year for each receiver in their households. Clearly, there were exceptions from the rules such as children living with their parents and not having their own income did not have to pay in addition. Also, usually only a single radio and TV receiver was counted per household. But already in the case someone owning a holiday flat equipped with radio and TV he or she had to pay twice.

The issue became trickier with the possibility to access broadcast content through the Internet. As a consequence, there were attempts to declare PCs and laptops as broadcasting receivers and demand the payment of a license fee for them as well. No doubt, in times when media consumption is taking place more and more on multifunctional portable devices such as smartphones and tablets or PCs/laptops such a model for a license fee does not seem to be sustainable anymore. Some countries like Germany have overcome this problem in the meantime by collecting the license fee in terms of one fee per household irrespective of the number of broadcast receivers, IP-enabled devices, or number of residents.

Even though a household based model for a license fee seems to be more up to date, the debate about the license fee is still going on. There are many people claiming they are not interested in PSB content and therefore think they should not pay. However, this is a very delicate issue. Some people may decide they will never go to the opera and still there is public money spent on opera houses as part of the cultural identity of a country. It is a political decision to use public money for this and no one is in a position to not paying tax on the grounds that he or she is not making use of particular cultural infrastructure. The same applies to public service broadcasting.

7.2   Public Service Broadcasters' Requirements on Distribution

In order to fulfill their obligations with regard to coverage and reach of their programmes public service broadcasters use a variety of different means to deliver their services. This includes both broadcast and broadband platforms, to a wide range of different devices, from large TV sets, home and car radio receivers to personal computers, tablets, and smartphones.

Not all distribution methods are equally suitable for all services and not all user devices come with the same capabilities. This may result in some cases in poor user experience. Nevertheless, public service broadcasters have to make sure that their content is delivered to their users at a guaranteed quality of service. It has to be available throughout the envisaged coverage areas as laid down in corresponding regulatory conditions.

The increasing number of possibilities to consume broadcast content on different devices and in various environments imposes growing burdens on broadcasters. Even though many broadcasters basically try to follow the "we-have-to-be-on-all-devices-with-all-our-content" philosophy, they are becoming more and more aware of the fact that such an approach may not be sustainable in the mid- to long-term future. With an increasing number of different distribution options the distribution costs are piling up beyond control. Moreover, different distribution channels usually call for different formats. Already today, between 10 % and 20 % of the overall budget of a television production is due to reformatting for different distribution paths.

Whatever ways of distribution public service broadcasters choose to utilize, there is a number of requirements that stem from their public service remit and apply for all conceivable distribution options. These requirements address both technical and non-technical issues. Distribution means which are relevant for public service broadcasters shall

- allow provision of content and services free-to-air, i.e. no additional recurrent costs for users in addition to the license fee and one-time costs for receiving devices and installations;
- allow delivery of content and services to the public without blocking or filtering, i.e. no gatekeeping;
- prevent any degradation to the content and service integrity, i.e. no modification of PSB content or service by third parties. Broadcast content and additional services (e.g., subtitles, HbbTV user interface) must be displayed unaltered and without unauthorized overlays;
- not be subject to discrimination compared to other equivalent services;
- allow the quality of service (QoS) to be defined by the public service broadcasters, including service availability, robustness, and reliability;
- ensure that the QoS is independent of the size of the concurrent audience;

- provide the possibility to tailor the geographical availability of the service according to the needs or regulatory obligations of the PSB, i.e. national, regional, or local coverage areas shall be possible;
- ensure that a distribution option supports at least a minimum number of services (e.g., a minimum number of programmes) in order to constitute an attractive platform for users;
- enable easy, straightforward access and usability;
- support prominence and findability of public service broadcaster's service offers;
- support low barriers for access to PSB content and services for people with disabilities (e.g., subtitles, audio description, and signing);
- ensure the ability to reach audiences in emergency situations; and last but not least
- shall be affordable under the conditions and constraints public service broadcasters have to operate.

These requirements are in general met concurrently by terrestrial broadcasting networks, in particular when operated by broadcasters themselves. On other distributions platforms usually only a subset of these requirements can be met at the same time. Broadcast distribution typically conforms better with this list of requirements than broadband distribution.

When having a closer look at broadband distribution a number of potential risks for public service broadcasters can be identified. Gate-keeping is one of the most important issues broadcasters have to face and cope with in a broadband environment. However, gate-keeping does not come in the same form everywhere.

The most common type of gate-keeping relates to access to distribution networks and platforms. This includes physical access to networks and access to online platforms on which audio-visual content of various forms is offered. Apparently, money is always a lever to overcome gate-keeping both on networks and on online platforms. However, the sheer number of all these commercially operated networks and platforms poses a real financial challenge for PSBs.

Furthermore, manufacturers of smart TVs have emerged as a special form of gate-keepers in recent years. Smart TVs offer access to broadcast distribution networks and broadband networks at the same time. This allows enjoying hybrid broadcast content offers such as HbbTV which combines linear TV with nonlinear content.

In the past switching on a TV set meant a certain linear TV channel was displayed. With contemporary smart TVs this is no longer necessarily the case. Smart TV manufacturers offer their own app based portals through which access to a variety of on-demand services can be established. Therefore, many manufacturers configure their smart TVs such that their portals are displayed on start-up and not any linear TV programmes. Clearly, all this can be reconfigured but there are a certain percentage of users which will not do that either because they are not able to or too lazy to engage and just do not care. This leads to a situation where the users are enticed away from the content offer of broadcasters at the very entry point, i.e. at the moment when they switch on the TV.

As broadcasters are interested in nonlinear content as well they certainly try to get their apps integrated into the portals of manufacturers. However, this means negotiating with different manufacturers independently about the position of the apps within the portal. It should come as no surprise that this will give rise to certain costs.

Similar issues can be observed with respect to search engines where obviously the position in search results is determined by the search algorithms. Whether or not this applies to major search engines one could imagine that the ranking of search results could be a very lucrative business. This carries over to app stores, third-party programme guides or recommendations offered to users. The positioning of products in this online environment is always also due to commercial agreements with the operators. As a matter of fact this has a significant impact on the findability of PSB content offers.

In many broadcasting companies slogans such as "We need to be where our customers already are" or "We have to be on all devices, with all content—linear and nonlinear—anytime, anywhere, at affordable costs and under the regulatory obligations and conditions we have to adhere to" constitute the strategic baseline along which decisions about content distribution are taken. This has led to a huge fragmentation entailing more and more costs. It seems that applying the requirements on distribution listed above to the available distribution options can help to better position broadcasters with respect to more efficient distribution and confine distribution costs.

Chapter 8
Strategic Considerations

At the beginning of broadcasting, distribution was an essential part of what broadcasting companies had to deal with. There were no independent network operators and hence broadcasters started to build and operate their own terrestrial broadcasting networks to get their radio and television programmes to their listeners and viewers.

Those days are long gone. The world has changed completely. Most broadcasting companies have given up or outsourced their terrestrial network operation in the meantime. They entered into business arrangements with corresponding network operators instead. Furthermore, with the proliferation of broadband networks new ways of distributing content to users have been unlocked.

As a consequence, the content distribution sector is getting more and more fragmented and complex. Therefore, it is vital for broadcasters to understand the dynamics of these new markets and develop a clear distribution strategy. Any such strategy must take into consideration

- the variety of services to be offered to users, i.e. linear, nonlinear, social media and combinations thereof;
- the devices people are using and the respective environments in which media consumption takes place;
- the regulatory conditions which in particular public service broadcasters have to adhere to;
- the economic aspects associated with content distribution; and
- last but not least, the capabilities of the distribution options.

As all of these aspects are mutually related, it is not constructive to address them individually. Rather, concepts are needed which allow separating those elements which are crucial for broadcasters from those which are of secondary importance. Some ideas and developments are introduced here without claiming that they would be sufficient to derive a solid strategy for future content distribution. However, they may indeed help broadcasters to shed light on some of the issues.

© Springer International Publishing Switzerland 2017
R. Beutler, *Evolution of Broadcast Content Distribution*,
DOI 10.1007/978-3-319-45973-8_8

8.1 Basic Observations

A simplified model of a typical company identifies three basic pillars on which the business rests. Depending on the actual organization they may be more or less independent from each other. Usually, there is a part of the company which is to produce or provide a product or a service. Another pillar of the business is administration. This encompasses managing people, resources, real estate, licenses, and rights. Finally, there is a part of the business which takes care of getting the product or the service to the customers under circumstances and conditions which are defined by the company and need to be under control in order to guarantee the quality of service. In all three areas of business activity there may be operational and strategic tasks to be fulfilled. Figure 8.1 sketches this idea.

From this perspective it is obvious that distribution is of vital importance. It is actually an integral part of any business. Unfortunately, in the domain of public service broadcasting this insight is not commonly shared. There are still broadcasters which consider distribution as of secondary importance. This probably traces back to the pre-Internet era when the distribution market was very stable and dominated by broadcasting networks. There were no alternatives and hence distribution of broadcast content was perceived as a non-issue. One could get the impression that this way of looking at distribution is still the prevailing attitude in the broadcasting sector.

Transmission capacity used to be limited on terrestrial broadcasting networks thereby containing competition. This put public service broadcasters into a pretty comfortable position for a long time, in particular as they were usually safeguarded by corresponding supportive regulation on top of that. Under such conditions, content and services of public services broadcasters were always highly appreciated by the audience.

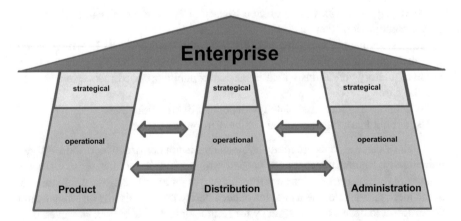

Fig. 8.1 Three pillar model of an enterprise

However, towards the end of the twentieth century more and more broadcasting companies started to outsource their network operating departments to cope with the increasing pressure to reduce costs. This gave rise to the creation of a number of broadcast network operators who took over the job from the broadcasting companies.

This arrangement was quite successful until with the advent of broadband distribution the rules of the game altered. Consumer behavior started to change dramatically. Broadcasters had to adapt by offering new services, in particular a whole universe of nonlinear services based on interactivity was developed.

By definition, classical broadcasting networks cannot provide a return channel and therefore broadcasters were starting to employ broadband distribution means. As this is clearly beyond the core expertise of broadcasting companies, new partners had to be found. Consequently, the already existing separation between production and distribution within the broadcasting companies became even larger.

The change of the distribution markets with the appearance of powerful broadband players often operating on a pan-European or even global level raises severe issues for broadcasters. There is no doubt that broadcasting companies have to develop a clear picture about how they would like to or need to distribute all their content and services in 5, 10, or 20 years, under what conditions, across which technical distribution paths, targeting what kinds of receiving devices and subject to what regulatory and economic constraints.

Finding guidance to cope with this challenge calls for the development of a sustainable holistic distribution strategy for broadcast content. This should take into consideration several aspects and concepts which are addressed in the following.

8.2 Use Cases

Without any doubt, the Internet opened the door to a completely new world of audio-visual services. Clearly, broadcast content is an essential part of this new world. However, what is a huge opportunity on one side may turn into a nightmare on the other side. In the past broadcasting companies have been those who determined the direction of the technological development in the broadcasting sector. They were in control of the process. This has changed today. Quite often it seems broadcasters are now driven by forces beyond their reach.

Slogans conveyed by marketing departments of the broadcasting companies such as "We need to be where our customers already are" unveil the problem. It seems that users are focusing on technologies and services which are developed outside the broadcasting hemisphere. If broadcasters do not want to lose their audience they have to follow. However, they are no longer making the pace nor giving the direction.

For a long time content production and service aggregation only had to deal with linear radio and TV programmes. Consequently, distribution technology only had to offer solutions to carry linear content to viewers and listeners. Certainly, there was progress with regard to the quality of the linear programmes. While starting with black-and-white pictures in an analogue environment today digital broadcasting technologies can offer enough capacity to even carry UHDTV content.

Broadband networks introduced interactivity. This means that users obtained a certain level of independence based on which corresponding demands emerged. Broadcasting companies had to deal with these demands. Their standard offer of linear programmes, i.e. programmes where an editorial department of a broadcasting company decides about the composition and the schedule, had to be complemented by new types of services. These were meant to give users the freedom to decide themselves when, where, and on which device the content is consumed.

Today broadcasting companies are confronted with the issue of deciding what types of services they have to offer, i.e. how much linear content versus nonlinear offers, what kind of nonlinear services and to which extent make use of existing and newly emerging Internet platforms. On the other hand, people are using more and more different types of electronic devices which are enabled to receive and display audio-visual content. This relates in particular to smartphones and tablets. Indeed, smartphones have turned into some kind of electronic interface to the world. Many people seem to be inseparably tied to their smartphone nowadays and therefore, they expect them to be able to satisfy every communication and information need. This includes access to any kind of broadcast content as well.

In contrast to the early days of broadcasting content consumption is no longer confined to the domestic living room. Mobile societies as we see them today come with a demand for access to audio-visual content while being on the move. This covers both the case where people are not at home but still are more or less static in one location as well as true mobile consumption in cars or public transport. Again, smartphones and tablets play an important role in this context.

Between content on one side and devices together with usage environments on the other side sits distribution. Before being able to decide which way content shall be delivered to users in the future broadcasters need to have a clear understanding what kinds of services they want or need to offer and which devices and usage scenarios they have to target. In that respect answering the question about the most suitable distribution options is the second step to be taken.

This insight was the starting point of an activity the European Broadcasting Union (EBU) initiated around 2012. The idea was to investigate available and future distribution options in a holistic manner putting the emphasis on the relation between content, devices, and environments in the first place. To this end, the concept of use cases was introduced. A variety of use cases were defined and categorized according to their relevance to broadcasters. Then, an analysis was carried out to identify distribution means which would allow enabling the relevant use cases.

According to the EBU methodology a use case is defined by three elements. Those are a broadcast service, a device, and an environment in which consumption takes place. Starting with this definition EBU carried out an extensive investigation on how relevant use cases can be enabled by existing and future distribution options. All details and results can be found in a corresponding EBU Technical Report [EBU14b].

As it turns out it is more or less straightforward to define a set of services which may be relevant for broadcasters. Certainly, linear radio and TV must be considered but also formats such as on-demand TV or podcasts are becoming more and more important. Finding a set of relevant user devices is also not very difficult. This may range from large screen TV sets and dedicated portable broadcast receivers to smartphones and tablets.

However, when it comes to the environments the issue becomes a bit more complicated. Delivery of broadcast content may be envisaged for outdoor or indoor reception, it may also be fixed, portable or mobile. From a user point of view this may all be not so important because what has probably more relevance is the question if a user is in control of the different possibilities to access content. At home this is certainly the case. However, while being on the move users have to rely on whatever type of network may be at their disposal.

Therefore, the approach developed in [EBU14b] introduces only two different environments which are utilized to define a use case. These are called "permanent" and "transient" in order to lay the focus on the level of control a user has on the availability of a distribution technology.

Permanent refers to a situation at home in the first place. This is evident. In their own houses or flats it is the users' own privilege to install a terrestrial antenna or a satellite dish if that is possible. Furthermore, they can get themselves a broadband connection with high enough data rates if they want. In an office the situation may be quite similar. Employees may be in a position to put up a portable radio or TV receiver if this is allowed by the employer as matter of principle.

On the contrary, a transient environment encompasses all situations where a user is on the move and therefore not in control of the access networks. This can be while driving in a car or using public transport means such as trains or buses. But it may also refer to a user waiting at an airport or a train station. Under such circumstances they do not have any influence on what kind of networks may be available nor on their capabilities. Hence, they have to use what they get, for example a slow WiFi network. There may be terrestrial broadcasting signals receivable but it may only be an FM signal and no TV at all.

It is clear that this classification can be debated in length. Depending on where to put the focus one may use additional and more refined environments to identify relevant use cases. Nevertheless, the approach setup by the EBU was the first attempt to structure the relation between content offered by broadcasters and users employing different devices in different environments. Table 8.1 shows different services, devices, and environments which can be used to form use cases [EBU14b]. However, it has to be emphasized that this list is by no means conclusive and can be adapted to the needs of individual broadcasters.

The combination of a service, a device, and an environment as listed in Table 8.1 allows creating a large number of use cases. In order to identify the most relevant use cases broadcasters may take into consideration current trends and expectations [EBU14b]:

Table 8.1 Services, devices and environments from which use cases can be derived

Services	Devices	Environments
Linear TV	Stationary TV set	Permanent
On-demand TV	Portable TV set	Transient
Catch-up TV	TV set in vehicle	
Linear radio	HiFi system at home	
On-demand radio	Portable radio	
Radio podcasts	Radio in vehicle	
EPG	Desktop PC	
Social media	Laptop Smartphone Tablet	

- Linear viewing is the primary way of watching TV content. There is no evidence that this will significantly change in the foreseeable future. In addition, time-shifted and on-demand consumption will continue to grow, but this may not weaken the linear viewing share.
- Migration of TV services from SDTV to HDTV will continue. More content will be offered as well, in particular with the introduction of new HDTV services. Furthermore, UHDTV will be introduced and may become the primary format in the future.
- Smartphones and tablets are increasingly used to access media services. This includes also access to broadcast content. Tablets are likely to replace traditional portable TV receivers.
- The majority of TV viewing, both linear and nonlinear, occurs in the home. Nevertheless, usage in transient environments will become increasingly significant.
- Social media are becoming more and more important for broadcasters in order to get feedback from audiences and to engage them.
- Hybrid broadcast-broadband services are becoming commonplace as sufficiently fast broadband access is spreading.

This list is certainly not conclusive. Broadcasters operating in their respective markets may have their own set of important trends. However, based on such observations there are certainly some use cases which are relevant to most broadcasters such as

- linear TV—large screen TV set—permanent environment (e.g., at home in the living room);
- linear radio—car radio receiver—transient environment (e.g., driving in a car);
- on-demand TV—tablet—transient environment (e.g., departure gate of an airport); or
- data service (e.g., Facebook site of a broadcaster)—smartphone—transient environment (e.g., in a public place).

Once a subset of highly relevant use cases has been identified the question about how to enable them in terms of distribution technologies can be addressed.

The EBU approach of defining relevant use cases is certainly no strictly scientific methodology. Rather, it is a subjective way to structure and prioritize the different areas of interest of broadcasters. It can thus constitute a first step on the way to developing a distribution strategy adapted to new services, technologies and changing market conditions.

8.3   Usage Patterns

Use cases proved to be a very useful tool in order to structure the complex interdependencies of the different elements of the broadcast value chain (see Fig. 2.1). Furthermore, they allow better understanding and visualizing the primary interests of broadcasters.

However, when looking at user's behavior it becomes evident that this is not sufficient to fully grasp the demands broadcasters are confronted with. Use cases usually do not come in isolation. It is true that there are times when users just sit in front of the TV screen and watch a movie for 2 h. But there are also more and more situations where people are employing different devices to enjoy services in parallel. Some are checking their emails or sending text messages to friends while watching TV.

Multi-tasking of this kind is without doubt a trend which some people are having mixed feelings about. Doing something else while watching TV may not cast a very positive light on the quality of TV content, it is however less disturbing than the bad habit of many people who are fiddling around with their smartphones while having a conversation with someone.

However, obviously not all parallel activities are bad. What is definitely interesting from a broadcaster's point of view is when people access a dedicated supplementary source of content to complement what they see on the screen by additional information. In principle, this kind of behavior can be described by a combination of different use cases for which the term "usage pattern" has been coined in [EBU14b].

Usage patterns correspond, for example, to a transition from one use case to another thereby generating a usage sequence in space and time. This implies a situation where there is a change of the service consumed, a change of the receiving device or a change of the distribution technology by which broadcast content is received. Figure 8.2 shows a particular example.

Another type of usage pattern comprises concurrent use cases. This means consuming one service on a given device while enjoying another service on another device. The services may not be linked at all or they may be associated in which case one can talk about a hybrid service. Figure 8.3 sketches a very common usage pattern involving a large screen TV set and a tablet.

Enabling distribution to support usage patterns may turn out to be very sophisticated as different distribution networks are usually involved. This may require cooperation between different network operators. Clearly, this can indeed be very

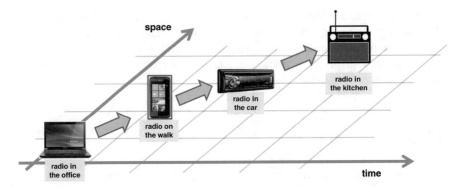

Fig. 8.2 Usage pattern representing a temporal and spatial change from one use case to another, i.e. listening to radio on different devices while moving from the office to home

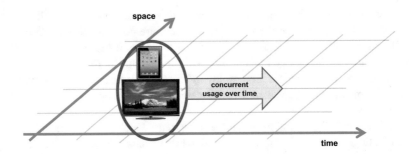

Fig. 8.3 Usage pattern representing a combination of two use cases over time while remaining in one location. The tablet may be used as a second screen device to enhance the linear TV programme or may be used independently

complicated to achieve as long as not all involved networks are under the control of a single operator. But even then there may be challenging technical issues. For sure, hand-over between different types of distribution technologies would be required and hence synchronization between the corresponding networks becomes crucial.

Cooperation of networks has been discussed under the label "convergence" now for quite some time in several fora in Europe such as the CEPT Task Group TG6 [ECC14] or the process of the so-called High-Level-Group leading to the publication of the Lamy-Report [EuC14a]. Broadcast-broadband convergence has been pushed very much by European regulators supported by parts of the mobile industry. In this context the European Commission issued a study on convergence. However, the study report concluded that the case for moving to a converged platform is not yet made [EuC14b].

From a purely technical point of view a system which would encompass both broadcast and broadband characteristics may be suited to cope with content delivery enabling usage patterns. However, no satisfying solution has been put forward so far neither in terms of feasibility nor with regard to supported business models

[EuC14b]. It remains to be seen to what extent developments such as eMBMS enhancements in 3GPP (see Sect. 6.2) or the 5G hype will eventually bring new insight (see Sect. 6.3).

8.4 Analysis of the Broadcast Value Chain

Defining use cases and usage patterns is certainly a good first step to analyze and structure the available distribution options. Clearly, the ultimate objective is to identify those distribution mechanisms and technologies which can enable the relevant use cases. Currently, there are five different distribution possibilities for broadcast content which are sketched in Fig. 8.4.

However, distribution of broadcast content cannot be considered independent of other elements along the entire value chain of the broadcasting sector (see Fig. 2.1). There are several additional constraints which have to be taken into account such as the regulatory regime, economic considerations, and aspects of globalized markets.

To this end, the EBU started to work on a general approach to understand the interrelations and deadlocks along the whole broadcast value chain which may support or impair the usage of a particular distribution option. This work was inspired by a method developed by TNO to analyze the case of web services [TNO14].

As a first step a simple canvas has been developed which should allow visualizing the value chain by highlighting the involved companies and organizations. This has been achieved by identifying seven distinct parts of the value chain. The canvas can be used to depict for a given distribution situation the flow of information and cash from one element of the value chain to others. Also, bottlenecks introduced by regulation or limited capacity of a given distribution option can be highlighted. Figure 8.5 shows the canvas for the broadcast value chain according to EBU [Rat16].

There is a high degree of freedom to define the different elements of the canvas. However, the layout shown here proved to be a good starting point for a more

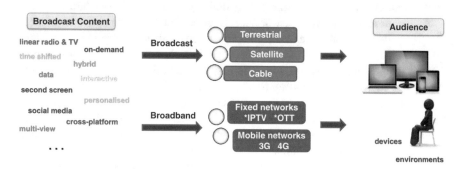

Fig. 8.4 Currently available distribution options for broadcast content

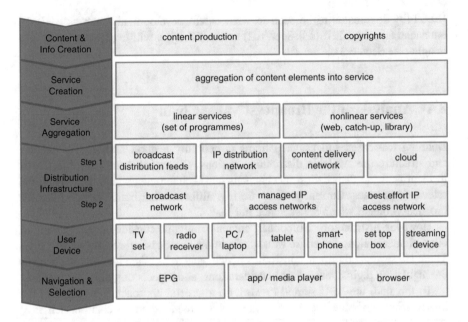

Fig. 8.5 Canvas of the broadcast value chain

detailed analysis. It captures all areas which have to be taken into consideration when trying to understand the interdependencies of broadcast content distribution.

The value chain starts with content and information creation. This covers all types of different formats broadcasting companies are currently offering. They are ranging from classical radio, TV programmes and all elements necessary for websites such as text, graphics, and metadata to even interactive content such as games. At this stage copyright issues have to be resolved already. Usually, copyrights not only refer to the ability to offer certain pieces of content but they also come with requirements about service integrity which have to be ensured all along the entire value chain. Furthermore, copyrights are very closely related to coverage areas. Most third-party content can only be offered within clearly defined geographical boundaries.

At the next step a broadcast service is generated from the different content elements. It is important to note that most broadcasting companies nowadays offer services which are composed from content they produce themselves plus content which comes from different sources, for example international production companies such as Hollywood studios.

Service aggregation refers to the act of bundling several services into a package which is offered to customers as a whole. A typical appearance of an aggregated service is, for example, a multiplex on a digital terrestrial television network which contains only 24h/7d programmes of the same broadcasting company. Similarly, bundling many different nonlinear offers in terms of an online library corresponds to another example of service aggregation. Aggregation of services is not a mandatory element of the value chain. There are, for example, many local radio stations which offer just a single radio programme, i.e. just a single service instead of a package of services.

After services have been assembled they are meant to be distributed to the users. Distribution can be divided into two distinct steps. In a first step the broadcast content has to be delivered from the studio or a play-out center to a handover point of a dedicated distribution network. In the case of a satellite network this is the satellite uplink station, while for cable or IPTV distribution these are the respective head-end stations. For digital terrestrial broadcasting networks the entry point is usually where the programme multiplex is created. Typically point-to-point connections are utilized to pass the content to these network entry points. This can be either accomplished by fixed fiber connections or wireless links.

In the case of broadband distribution the first step of distribution corresponds to making content available to an Internet provider who will offer it through dedicated websites. Passing content to a CDN in order to cache content closer to users is another example for this.

The second step of the distribution element in the value chain canvas is what is usually associated with the term distribution. It relates to employing some network, either broadcasting or broadband, to actively bring content to people.

Broadcast content can be consumed on many different devices today. However, depending on which device is used and under which conditions there may be completely different implications in terms of efficiency and costs. Therefore, it is important to introduce this field into the analysis of the value chain.

The final element in the value chain analysis is navigation and selection. Depending on the various ways of accessing content there are also very different ways of navigating through the offer. As this is happening exactly at the point where users are taking decisions navigation and selection processes have become crucial for broadcasters. In particular, with the increasing influence of gatekeepers of various kinds it is vital for broadcasters that their content can be easily found and accessed.

A typical application of the value chain canvas would be to visualize all involved parties in the corresponding fields of the canvas for a given use case. Then, the flow of content can be visualized together with the cash flow, for example. Particularly important bottlenecks can be highlighted in the graph which may indicate important issues.

In order to give an example Fig. 8.6 shows the simple case of linear TV distributed over a terrestrial broadcasting network. This is not a real-world example. Rather, it is just a notional visualization highlighting some of the interrelations between different players. Clearly, the level of detail can be increased to fit any desired granularity of information. In principle, any relevant information about the regulatory regime or potential gatekeepers can be introduced as well.

Even though the canvas put forward in Fig. 8.6 may almost appear trivial, things can become very complex very easily if more than one very delimited situation is considered. Then, the value chain canvas may indeed become a viable tool to keep track of all issues. Under such conditions this approach could be particularly useful

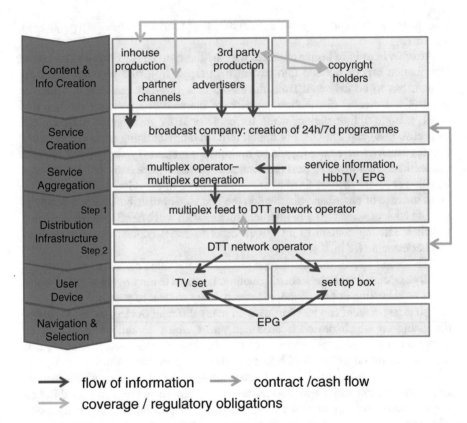

Fig. 8.6 Application of the value chain canvas to the case of distribution of linear radio or TV programmes over a digital terrestrial network

to analyze a specific local market with concrete market players and their mutual relations, commercial and service agreements, applicable regulation, etc.

Furthermore, generating a value chain canvas may be in particular helpful when it comes to comparing distribution of the same type of content, i.e. for example linear television programmes, over different distribution paths. Offering linear TV as a live stream has certainly different implications than making use of a traditional terrestrial broadcasting network.

8.5   Business Arrangements

It should be obvious that distribution of audio-visual content does not come for free. However, there is a huge variety of different arrangements with respect to how distribution is enabled and paid for. Content distribution is the link between content generation on one side and content consumption on the other side. To this end, dedicated distribution networks need to be employed. Building and operating distribution networks requires huge investments and usually entails large operational costs. Any content provider making use of a dedicated distribution network will have to pay in one way or the other for content carriage.

In principle, there are two different sources of revenue for network operators. The first source are the content providers who want to make sure that their products arrive at their customers in a clearly defined way and under the regulatory and economic conditions they have to adhere to. The other source of income is the customers of the network operator who seek access to various types of audio-visual content.

However, in the broadcasting sector special conditions apply. This is due to the existence of public service media companies. They have been created by means of political decision processes in the different countries. The rules public broadcasting companies have to adhere to may vary from country to country but PSBs are everywhere funded by public money. This can be raised in terms of a special tax or every household has to pay a dedicated fee every year which is used to support all aspects of public broadcasting from content creation to distribution.

In addition to the particular funding of public service media there is usually special regulatory conditions placed on them. Ubiquitous coverage and free-to-air provision of content are among the most important tasks. In some countries advertisement is allowed in PSB programmes, in other countries this is left to commercial broadcasters only.

Quite often additional regulation applies for the distribution of public service content. Most common are the so-called must-carry rules. In order to make public service programmes available throughout a given country regulators may impose that network operators have to offer a certain set of PSB programmes as part of the content packages. Must-carry rules are meant to regulate proper distribution of PSB programmes (see also Sect. 4.2.1). In many countries must-carry rules are applied to cable networks in the first place.

A part of the budget of public service broadcasting is dedicated to distribution. Some PSBs can autonomously decide how to use the available money. Others are given a well-defined amount of money for selected distribution options. The latter case means shifting investments from one distribution form to another usually requires a lengthy administrative process. In times when conditions are changing quickly as it is the case since the Internet and broadband connections have become available, this may put PSBs in a difficult position.

As a consequence of all these special conditions, public service broadcasters are neither normal market participants nor are they standing aside the market dynamics like an independent observer. For commercial broadcasters the case seems to be clearer. They are entirely governed by economic forces. The one and only objective is to make profit. Therefore, any decision to make use of one or the other form of content distribution has always to be justified by economic reasons only.

Taking into account all the different conditions and constraints under which PSBs have to operate there are basically five different models under which business arrangements between public service broadcasters and network operators could be addressed.

1. **Network operation by broadcasters**

   This has been the standard way of distributing content to users in the pre-Internet era when broadcast content consisted in linear programmes only. Historically, broadcasters deployed and operated their own terrestrial networks, at the beginning analogue, later digital terrestrial networks. In a number of countries, PSBs still operate their own terrestrial transmission networks, although this model is nowadays less prevalent in Europe.

   With respect to distribution of broadcast content over broadband networks, this model would imply that a broadcasting company would generate broadcast services and operate a broadband network end-to-end, i.e. from the play-out center of the broadcasting company to the users wherever they are seeking access to the content.

   Considering the practical hurdles to be taken and the huge investments it seems unrealistic that broadcasters would eventually enter into operation of their own national or at least regional broadband network.

   The costs for operating their own networks, whether broadcasting or broadband have to be borne by the broadcasting companies. PSBs have to spend parts of their license fee based budgets on network operation. Commercial broadcasters could roll-off network investment and operational costs onto the users in terms of subscription models to access content.

   It may actually be a trivial statement but network operation by broadcasting companies themselves does not call for direct business arrangements between broadcasters and network operators.

2. **Distribution as a service**

   Historically, this was the next natural step which followed the decision of broadcasters outsourcing their own network operation. Distribution as a service means that broadcasters commission a network operator to carry out the distribution of the broadcast content by a dedicated network.

   The conditions and constraints for the distribution are laid down in a contract between broadcaster and network operator according to which the operator has to plan and design the network. From a PSB point of view these conditions refer primarily to the envisaged coverage areas, the quality of service, and the number of programmes to be distributed. These parameters are defined by the broadcaster on the basis of given regulatory obligations.

   Traditionally, the distribution as a service approach has been widely applied for distribution over broadcasting network, primarily terrestrial and satellite networks. However, the current discussion about using eMBMS LTE networks to deliver in particular linear radio and TV programmes to smartphones and tablets (see Sects. 6.2.1 and 6.2.2) came across distribution as a service as a potential option to define a business arrangement between a broadcaster and a mobile network operator.

Broadcast content providers have to pay the network operator for distribution of their content. Depending on the circumstances users may have to pay as well. This depends on the market situation. Therefore, different variants are possible in principle.

3. **Must-carry distribution**

In many countries national regulators can impose obligations for the network operators to provide services to the users, for example public broadcasting services. In the EU, the telecom regulatory framework [EuC16a] supports must-carry rules for some specific audio-visual services if the distribution networks over which they are offered are the way to receive these programmes for a significant number of consumers.

Must-carry rules can be applied to distribution options where subscription to a service is a prerequisite to get access to content. In the past, must-carry obligations were applied primarily to cable/satellite TV providers with a significant market share. The must-carry rules usually stipulate the inclusion of the PSB channels in their basic content offers so that they are available to all subscribers.

Application of must-carry rules does not necessarily mean that distribution is for free. Depending on the specific regulatory conditions, the scope of the obligations or the kind of service, different models exist. In some cases, network operators pay for the content they make available through their networks, in other cases the PSM service providers pay for carriage.

Must-carry rules have been developed in the context of delivery over broadcasting networks. In principle, they could be adapted to broadband distribution as well. This refers in particular to managed services which are subscription based and offer access to a bundle of linear programmes or nonlinear content offers. In this case, must-carry rules could simply impose the inclusion of PSB content in the portfolio of the managed service provider in the same way as in broadcasting networks.

4. **Partnership based distribution**

The partnership model for distribution of broadcast content relates to a situation where a network operator is offering access to content in addition to connectivity. The network operator offers, for example, access to the Internet and in addition bundles a set of audio-visual services stemming from different sources into a different packages. Access to these packages is usually enabled by means of subscription.

Under these conditions a broadcaster may decide to place his programmes to the disposal of a network operator who incorporates them into his content packages.

There are several ways of organizing the business relation between content provider and network operator. The primary objective of the network operator is to monetize the content he can offer. To this end, the operator has a contractual relation with his subscribers whereby revenue is generated. In some cases, the operator has to pay the content provider to integrate particular content into his offer. In other cases, content providers pay the operator to get their content distributed as part of the package. In any case, the distribution costs are to be borne by the network operator.

This type of business arrangement between a broadcaster and an operator can also be applied to Internet platforms offering audio-visual content (see Fig. 2.2). In this case the network operator is a platform operator who does not have his own network infrastructure. Platforms such as Netflix [NeF16], Hulu [Hul16], Maxdome [Max16], etc. fall into this category.

5. **Over-the-top distribution**

Over-the-top distribution did not exist in the traditional broadcasting ecosystem. It developed in connection with the way access to the Internet is provided. In order to make their content available content providers need to have a business relation with an ISP to make their content available on the Internet. Users who would like to consume this content also need to have access to the Internet. To this end, they need to subscribe to an ISP providing them a corresponding broadband connection. This can be offered through either fixed or mobile broadband networks. The data requested from the web server with the broadcast content is routed through the open Internet.

In the case of the OTT distribution model there is only a business arrangement between the content provider and his ISP providing the web server. There is no business relation with any other entity engaged in routing the audio-visual content to the user. On the other side, the user is a customer of his ISP to which he has to pay corresponding access fees.

In order to ensure better quality of service and to limit the shortcomings of the best-effort open Internet distribution broadcasters are employing CDNs (see Sect. 3.2.5). The costs for these have to be borne by the broadcasters. This does not change the business relation between users and the IPS they are subscribed to.

The OTT model allows the broadcast service providers to address potentially large, indeed global, audiences without high upfront investments in broadband infrastructure.

In practice, the lines between these different business arrangements or distribution models are blurred. In particular, the case of platform operators as discussed under the partnership model is quite closely related to the OTT model. As platform operators usually do not operate their own broadband networks and in addition do not engage with broadband network operators directly, the distribution of content in their offers actually uses best-effort open Internet delivery mechanisms.

In particular in relation to the development of new distribution technologies which may also be interesting for broadcasting companies the question about the appropriate business arrangement between broadcasters and network operator is of fundamental importance. In the mobile broadband world revenues are generated through direct business relations between users and MNOs. Many different subscription packages are offered to customers.

At the moment when the idea of free-to-air distribution of audio-visual content becomes an issue this concept can no longer apply. It is important to understand though that FTA distribution of public service broadcast content does not mean that network operators are forced to carry content over their networks without getting paid. Rather, business arrangements between public service broadcasters and network operators will replace the traditional relation between network operator and user.

The discussion in 3GPP about eMBMS enhancements revealed that actually the prospects of a potentially promising business arrangement between broadcasters and MNOs could be a much stronger driver to enable a corresponding technical specification than any technical arguments about better performance or higher efficiency.

8.6   Distribution Costs

Distribution of broadcast content is an expensive affair. A typical satellite transponder which can carry about 4 HDTV programmes costs several million Euros per year. Nationwide coverage with one DTT multiplex on a digital terrestrial network may incur even 10 times higher costs per year. Depending on the number of users which are utilizing a given distribution option the costs per capita may hence vary significantly.

Obviously, broadcasting technologies come with high fixed costs which are mainly due to the investment in the network infrastructure. Once the network is up and running additional users do not increase the distribution costs. Within the coverage area of a DTT network or within the footprint of a satellite beam the potential number of concurrent users is practically not limited.

This is completely different with broadband networks. Broadband networks as of today (2016), both fixed and mobile, rest on unicast connections to serve the users. This means that the more users are requesting content the more traffic is generated on the network. The more bytes are transferred the higher the bill will be for the broadcaster in the end.

If, for example, during an important sports event, i.e. the Olympic Games or a World Football Championship, users are connecting to the website of a broadcasting company in order to watch linear TV as a live stream, they are all served by individual unicast connections. From a broadcaster point of view this is certainly not efficient use of resources, let alone the associated distribution costs.

One way to resolve the issue with distribution costs on broadband networks would be to implement a multicast/broadcast mode and use it whenever there is a high concurrent demand for the same content. However, such a multicast/broadcast mode would need to be implemented end-to-end along the entire chain from the broadcaster to the user device. Since there are many different players involved along the line it seems to be almost hopeless to accomplish such a more efficient way of content delivery (see Sect. 3.2).

For the time being, the total costs for broadband delivery are still just a fraction of the expenses for broadcasting networks. The reason is that still the amount of broadband usage is almost marginal compared to consumption over broadcasting network, despite the hype about the Internet as the future of the broadcasting business. However, the trend is pretty clear. It must be expected that there will be a significant increase in broadcast content consumption over broadband networks and therefore the total costs are likely to rise dramatically.

Moreover, already now the costs for broadband distribution are significantly higher than those for broadcasting networks when put in relation to the viewing shares. Many investigations have been carried out in recent years to better understand the issue. The BBC Trust, for example, stated clearly in their 2013 report [BBC13]

> 3.67 While cost effective when compared to commercial benchmarks, online distribution (of the iPlayer only) delivers just over c.2 % of the BBC's TV viewing, while the costs associated with delivery of the on-demand service (core traffic-related costs only plus the YouView share, adjusted for iPlayer's share of total streams), is just under 12 % of the BBC's total distribution bill. By comparison, as noted above, traditional distribution costs make up 87 % of the total, but this delivers nearly 98 % of viewing.
> 3.68 Put another way, for each percentage point of viewing share delivered across linear and non-linear, linear distribution represents expenditure of £2m while the non-linear (iPlayer only) distribution expenditure per percentage point of viewing is c.£12m (six times as expensive). On this simple basis, the BBC's expenditure on linear services appears to be far better "value for money".

This shows that under the current operational conditions broadband distribution is disproportionately expensive. The only thing which keeps the costs within acceptable limits for the time being is the fact that broadband distribution represents a small share of the overall content distribution.

Discussions about costs of distribution over certain networks have been one of the hot topics in recent years. In particular, the question of distributing broadcast content over LTE networks triggered many heated discussions. Usually, these debates were initiated in the context of the battle about more spectrum for mobile services below 1 GHz.

In Europe over the last ten years the spectrum range between 790–862 MHz (the "800 MHz band") has been released from terrestrial broadcasting in order to be used by the mobile service. The same shall happen with the band 694–790 MHz (the "700 MHz band") until at latest 2022. Giving up both UHF bands put a lot of pressure on DTT in Europe. The question is if the remaining part will be sufficient to guarantee enough freedom to further develop the terrestrial broadcasting platform.

In this context some representatives of the mobile industry argued that the days of DTT are over anyway and its job could be done by LTE. In other words, the claim is that in the mid-term future DTT would be replaced by LTE as this would be more efficient and flexible. It is more than doubtful if this is indeed true but the real question is would broadcast content distribution over LTE networks on a large scale be affordable for broadcasters at all.

The EBU has set up a special project team in 2011 called CTN-Mobile,[1] to which the mobile industry has been invited. The idea was to tackle the question what can LTE do for broadcasting. It was decided not to talk about spectrum issues as it was clear that this would prevent any progress. Many technical issues were dealt with and two reports have been published which were written jointly by broadcasters and representatives of the mobile industry (see [EBU14a] and [EBU15b]). As such CTN-Mobile proved to be a very important forum to make progress on issues which are both in the interest of broadcasters and mobile industry. Even though there have been similar attempts to achieve the same it is probably safe to say CTN-Mobile was the most successful so far.

Unfortunately the call for participation was only followed by the manufacturers and vendors of the mobile industry. Mobile network operators did not join the activity. This would have been indeed beneficial in particular when CTN-Mobile started to tackle the issue of distribution costs of broadcast content over LTE networks.

Initially, the discussion revolved around comparing distribution over DVB-T2 networks with LTE networks. However, this is comparing apples and pears. DVB-T2 is a unidirectional broadcasting system while LTE could offer a broadcast mode in terms of eMBMS plus unicast capabilities. Hence, it is important to clearly define the use case. To this end, the first analyses were carried for delivering a set of linear HD programmes across an entire national territory at a given QoS.

Both broadcasters and mobile representatives carried out their own cost calculations for the distribution of linear programmes based on certain assumptions about the main cost driving factors [EBU14a]. It does not come as a surprise that the results were different by an order of magnitude. While broadcasters expect the investment and operational costs per year to be almost ten times as high as those of a DVB-T2 network, the mobile side claimed it could be done almost at the same price as DVB-T2.

At the moment it is not possible to arrive at a conclusive assessment. This may also be due to the fact that for the time being the idea of distributing broadcast content over mobile broadband networks does not seem to be attractive to MNOs. More discussion is needed about potential business arrangements between broadcasters and operators. In particular, more education is needed that free-to-air broadcasting does not mean that there are no revenues to be made and that MNOs have to carry public broadcasting services for free.

So far, the business model of MNOs is user based, i.e. revenues are generated through the user subscriptions. In the case of an MNO being commissioned to take care of the distribution of broadcast content quite likely the distribution as a service business model may be applied. Then, the existing B2C relation would be replaced by a B2B relation between broadcast company and mobile network operator.

---

[1]CTN was the name of the mother group and stands for "Cooperative Terrestrial Networks." The appendix "Mobile" is to indicate that the scope of this project team lies on mobile networks.

8.7   Assessment of Distribution Options

As has been discussed so far, the attempt of finding the right distribution means for broadcast content, be it linear or nonlinear, is an exercise which is far from being trivial. There are indeed several completely different aspects which need to be taken into consideration at the same time. Confronted with developments in the global audio-visual markets which are more and more fragmented no one should expect simple solutions.

With regard to distribution the overall picture is very confusing. Therefore, it is probably a good idea for broadcasting companies to recollect what their primary objectives are. As more and more broadcasters have outsourced their distribution departments and started to commission content distribution to dedicated network operators, it seems to be obvious that the core competence of broadcasting companies lies at service creation and aggregation.

Outsourcing distribution may be a legitimate solution in particular in response to the increasing pressure on broadcasting companies to reduce costs. However, if there are no in-house distribution departments anymore, it becomes vital to build a strong strategic position towards distribution issues. Only this way a broadcaster is able to understand where to focus on and keep control in relation to contracted external partners.

The approaches and methodologies discussed above may help to develop an understanding of the proper interests of broadcasting companies. While recollecting their core business broadcasters have to come up with an idea about what use cases are indeed relevant for them or will become relevant in the future. This includes in the first place what types of services they want to offer. Furthermore, the expectations and habits of users need to be taken into consideration. This means in which environments they consume content and which devices they are using. In other words, the first step is to define relevant use cases as explained in Sect. 8.2.

Once this is clear potential distribution technologies can be investigated to see if they could enable the relevant use cases. However, any promising distribution option which may be suitable to enable an important use case has to be crosschecked against the fundamental requirements of broadcasters. Obviously, at this point there is a difference between commercial and public service broadcasters. The latter have very particular requirements which are due to their special remit. This has been discussed in Sect. 7.2. The basic requirements of public service broadcasters are sketched again here in Fig. 8.7.

Figure 8.8 shall underline the central role of the requirements for the process of determining which are the best distribution options under given conditions.

There is no doubt that neither a single use case nor a single distribution option will be sufficient to fulfill the remit of public service broadcasters. Actually, the trend irresistibly is moving towards broadcasters having to be on all devices, in any location and at any time. Furthermore, this has to be accomplished under affordable conditions for both broadcasting companies and last but not least the users.

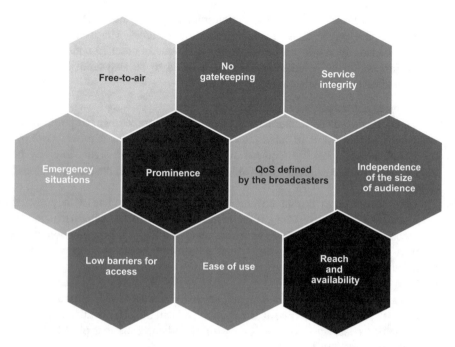

Fig. 8.7 The basic requirements of public service broadcasters

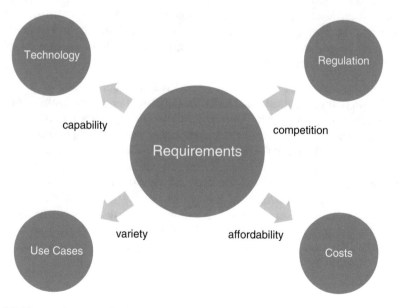

Fig. 8.8 The requirements of public service broadcasters in relation to external factors

The assessment of a distribution option with respect to its ability to enable a relevant use case may be subdivided into four different areas. Those can be entitled technical capability, reach, costs and market position. In order to understand if a distribution option is viable several questions need to be answered. In the four indicated areas the following issues are relevant:

- **Technical capability**

    Is the envisaged distribution technology capable for the intended purpose, for example to carry tens of linear HDTV programmes? Does it provide enough capacity and support the required quality of service? Can signal integrity be guaranteed? Is the technology scalable in order to cope with demand variations?

- **Reach**

    Is it possible to design coverage areas which are in line with potential regulatory obligations? Can different reception modes be supported such as outdoor, indoor, fixed, portable or mobile reception? Does the distribution option reach the targeted user devices? Is it possible to reach a concurrent mass audience?

- **Costs**

    Does a viable business arrangement between broadcaster and network operator exist? Are distribution costs transparent and can they be contained? How do distribution costs scale, e.g. with the size of audience, number of channels, or quality requirements?

- **Market position**

    How can the regulatory obligations of broadcasters, in of particular public service broadcasters be met? Can free-to-air distribution be enabled? Is there any gate-keeping and how can it be avoided? Can prominence and findability of content be guaranteed? Who are the competitors which are making use of the distribution technology as well?

Questions like these need to be considered in order to decide if a given distribution option is indeed suitable for the purpose of broadcasters. In addition to this, applying an analysis of the value chain as described in Sect. 8.4 for a given use case and a potentially suitable distribution option allows detecting potential areas of difficulties which may call for particular further consideration and analysis.

There is chance that these kinds of analyses will also indicate that none of the existing distribution options is viable to enable some of the relevant use cases. This can be due to technical deficits or it may turn out that the distribution costs will just not be bearable. Even though such a conclusion may not be satisfying in the first place it may nevertheless be very useful. It allows identifying aspects which need to be addressed and can therefore help to decide where to engage, i.e. in which organizations, technical development projects, or regulatory activities.

Chapter 9
Broadcasting in Uncharted Waters

Broadcasting has played a very important role in societies for decades. In particular public service broadcasting has contributed to the cultural diversity. The remit of PSBs to serve all social milieus with a huge range of different programme offers from news, sports, fiction, dedicated kids programmes, education to academic events has provided social cohesion to Europe for a long time. Pan-European content such as the European Song Contest or live transmissions of big sports events are still the most popular programmes. However, there is also a plethora of regional content targeting at the demands of particular areas or linguistic subsets of the population.

However, once societies started to enter into the digital era the picture changed. All of a sudden new content providers exploiting completely new technologies started to compete with incumbent broadcasting companies to win the public's favor. Equipped with huge budgets coming from international investors these new players put PSBs under high pressure.

Broadcasters felt themselves confronted with the need to develop new programme formats including an active engagement in social media. Many radio and TV shows of today would not be feasible without social media integration.

Not only content changed. There are also new devices which can be used to consume broadcast content. They are a blessing and a curse at the same time. On one side smartphones and tablets offer new opportunities for engaging with the audience. On the other hand, as has been discussed they are beyond the control of broadcasters at least in terms of offering all types of content there.

New devices and new forms of content call for new ways of distribution. No doubt there is progress on the broadcasting technology side. Hence, if broadcast distribution technology can guarantee to reach smartphones and tablets by whatever means this is certainly fine for broadcasters, if not they need to find other ways. All attempts to integrate a DTT receiver into smartphones and tablets have completely failed so far from a commercial point of view. There are technical solutions but both manufacturers and network operators did not show interest. Therefore, the only way forward may be to change existing mobile technology to open the door for broadcast

© Springer International Publishing Switzerland 2017
R. Beutler, *Evolution of Broadcast Content Distribution*,
DOI 10.1007/978-3-319-45973-8_9

content. Yet, it remains to be seen if the engagement of broadcasters in 3GPP to support their case will be successful in the long run.

The other big thing is certainly 5G. It is hyped everywhere, in politics, in regulation, in international spectrum management, and in industry anyway. For the time being there does not seem to be a clear understanding what 5G will actually be. Sometimes discussions almost get a religious touch as some stakeholders probably consider 5G to be a savior to all problems western societies are currently facing.

Nevertheless, what is clear today is that 5G will have an impact at least on distribution of broadcast content. The big question, however, will be what role 5G will play in that respect. Will it just be an additional distribution path on the broadband side, will it integrate all broadband networks under one roof or will it finally also cover the broadcasting distribution options? Is it about integration in the first place or is 5G set to replace all existing technologies by something completely new in the long run? In any case, it seems to be obvious that seamless interoperability of different technologies and networks is a prerequisite for 5G to be successful.

It has yet to be seen whether the high-flying promises and visions concerning 5G will come true or to which extent they may be realized. In organizations such as 3GPP or the WP5D of ITU-R the implicit assumption is that 5G is about wireless communication which is expected to revolutionize our world. This will go hand in hand with a plethora of new services and applications spreading out into every corner of societies around the globe. As a consequence, data traffic volumes to be carried over 5G networks will increase dramatically. But this means that backbone and backhaul infrastructure of future communication networks have to be adapted correspondingly to be able to cope with this traffic.

The consequence of this seems to be clear. Even though there are high-flying expectations with regard to new possibilities for portable and mobile communication the major investment and innovation induced by 5G will quite likely take place in wired infrastructure. This means that fiber networks have to become ubiquitous. The wireless 5G sector may then just play the role of providing access to the omnipresent fiber network infrastructure for portable and mobile devices.

If this is true, then the only question which is left is how large will the "last mile" actually be? Will there be fiber access points in every room in every house? Or just outside our houses around the corner down the road? At the end of the day it will be a question of investment and expected revenue as usual but wireless communication will be reduced to rather short distances compared to the current situation.

This raises a very important question. For making all 5G dreams come true it seems to be crucial to have a ubiquitous fiber network reaching out to every household, every factory, every office, and every public place. The question is who shall bear all the necessary investments? It seems to be very unlikely that such fiber penetration can be achieved by relying on market forces only. fiber roll-out may be commercially viable in urban areas but probably not in rural areas.

If, however, broadband connectivity is indeed so important for the future development of western societies as unanimously presented in all the 5G debates and discussions, then its availability is tantamount to that of water and electricity.

If this is right, then it is hard to avoid the question if provision of broadband connectivity is not a common task of communities or societies. This would suggest accomplishing fiber roll-out on the basis of public money. In times when every development is left to market forces this proposal may probably meet with no response. However, there are initiatives where small villages or cities started to engage themselves in this direction because the praised market forces simply fail.

In addition to the fiber issue there is another aspect of 5G which may stir up the current broadband ecosystem. Some of the developments in 5G such as autonomous driving will certainly build on device-to-device communication. Information will be sent directly from one device to another in real-time without routing through a remote base station. Without any doubt there is still a need for traditional base station type communication, for example, for those services which have to be made available for all vehicles such as detailed traffic information or road conditions. But it remains to be seen how far future communications will deviate from the classical approach still used today.

It seems to be evident that all these developments will have a dramatic impact on the role mobile network operators will play in the future. Their current business models may no longer be sustainable in such an environment. It can be expected that the primary business will be wire-based with a wireless extension depending on the circumstances.

Against this background it is natural that the incumbent stakeholders in the broadband business try to unlock new markets which could benefit from 5G technologies. This is the reason why in particular the European Commission is so eager to engage with the so-called verticals, i.e. independent vertical market sectors. Those are smart cities, e-health, energy, traffic, and media and entertainment.

Finding new customers in new markets is fine and is certainly welcome by most market participants. However, making business with new customers requires offering a product or service which suits the needs of the customers and is able to fulfill their requirements. Exactly this is the crux of the matter at least regarding the mobile network operators. At least for the time being it seems they did not understand the signs of the times. Their behavior in 3GPP shows that they want to get hold of new customers; however, they still believe their current business models will also work in the future. Quite likely they may be confronted with a different reality.

For content providers such as broadcasting companies the 5G developments will pose a completely new situation as well. The issues related to distribution network access and gate-keeping will become more pronounced. This will intensify the discussions about how to regulate distribution of audio-visual services. In particular, it will be crucial for public service broadcasters that they will not be victims of these developments which to a large extent will be determined by economic interests of the telecommunication industry in the first place.

The change of the distribution markets with the appearance of powerful broadband players often operating on a pan-European or even global level, calls for sound strategic analysis of distribution paths for broadcast content. Broadcasting companies have to develop a clear picture about how they would like to or need to distribute their different types of content and services in 5, 10, or 20 years, under what conditions, across which technical distribution paths, targeting what kinds of receiving devices and subject to what regulatory and economic constraints. Thus, the development of a sustainable distribution strategy seems to be required more than ever.

The lesson which can be learned from the engagement of broadcasters in the 3GPP and 5G processes is that direct engagement is a key to success. However, broadcasters in Europe and elsewhere are currently trying to cut costs on a large scale. This also has an impact on participation in technological developments, debates about new regulation and lobbying in relevant groups and organizations. This seems to be a step into the wrong direction. In fact, in times when broadcasters do no longer control distribution paths themselves it may become vital to change this position and put significantly more effort, both people and money, into these activities. It feels that this may be the only possibility to maintain the market positions and their role in society.

The engagement of broadcasters in the field of newly emerging technologies has to clearly articulate their concerns. Key technical, regulatory and economic issues have to be addressed, for example: Will there be enough capacity for PSB content on future distribution networks? What about control of the networks, who is operating them? Will future networks be able to allow PSBs meeting their regulatory obligations in terms of free-to-air access, coverage, QoS, number of programmes, regional content, etc.? Furthermore, the migration from today's distribution infrastructure has to be clarified, in particular with regards to costs. Quite likely, new business arrangements between broadcasters and network operators have to be envisaged. And finally, issues such as competition in globalized markets and gate-keeping have to be properly addressed by taking into account the public service broadcasting point of view.

Slogans such as

We need to be where our customers already are

or

We have to be on all devices, with all content—linear and nonlinear—anytime, anywhere, at affordable costs and under the regulatory obligations and conditions we have to adhere to

may give the right direction but they need to trigger action. Broadcasters have lived too long on their historical role as primary providers of radio and television programmes. Most of them believed at the beginning that even in an Internet centric world they would still keep their pole position. In the meantime this turned out to be a misjudgment. Actually, their prominent role in the past was due to the fact that they were the main players in a closed or at least limited distribution market. This safety net has definitely gone.

Even though forecasting is difficult, in particular when it relates to the future as Mark Twain once said, strategy managers of broadcasting companies, in particular the public service broadcasters, therefore have to take the right decisions now in order to be well positioned in 5, 10, or 15 years.

Bibliography

[AdS16] en.wikipedia.org/wiki/Adaptive_bitrate_streaming, 2016

[Alc11] Alcatel-Lucent, *Introduction to Content Delivery Networks*, www.cachelogic.com/images/documents/IntrotoContentDeliveryNetworks.pdf, 2011

[APT16] Asia-Pacific Telecommunity, www.aptsec.org, 2016

[ARD16] ARD, *ARD Mediathek*, www.ardmediathek.de/tv, 2016

[ARI05] Association of Radio Industries and Businesses (ARIB), *Transmission System for Digital Terrestrial Television Broadcasting*, ARIB STD – B31 Version 1.6-E2, 2005

[ASM16] Arab Spectrum Management Group, http://asmg.ae, 2016

[ATS07] Advanced Television Systems Committee, Inc., *ATSC Digital Television Standard, Parts 1 – 6*, A53, 2007

[ATS15] Advanced Television Systems Committee, Inc., Technology Group 3, atsc.org/subcommittees/technology-group-3, 2015

[ATS16] Advanced Television Systems Committee, http://atsc.org, 2016

[ATU16] African Telecommunications Union, http://atu-uat.org, 2016

[BaE14] Bar-El, H., *Protecting network neutrality: both important and hard*, www.hbarel.com/analysis/policy/what-is-network-neutrality?page=2, 2014

[BBC13] British Broadcasting Corporation Trust, *The BBC's distribution arrangements for its UK Public Services*, downloads.bbc.co.uk/bbctrust/assets/files/pdf/review_report_research/vfm/distribution.pdf, 2013

[BBC16] British Broadcasting Corporation, www.bbc.co.uk, 2016

[Beu04] Beutler, R., *Frequency Assignment and Network Planning for Digital Terrestrial Broadcasting Systems*, Springer, New York, 2004

[Beu08] Beutler, R., *Digital Terrestrial Broadcasting Networks*, Springer, New York, 2008

[Bjo15] Björkman, P., *Wireless Distribution in the Future Media Landscape: Broadcast-like Services*, ETSI-EBU workshop "Distribution beyond 2020", ETSI Headquarter, Sophia-Antipolis, www.etsi.org/past-workshops-and-events/presentations-wireless-media-distribution-beyond-2020, 2015

[Bun16] Bundesnetzagentur, www.bundesnetzagentur.de, 2016

[Cab16] en.wikipedia.org/wiki/Cable_Internet_access, 2016

[CDN16] en.wikipedia.org/wiki/Content_delivery_network, 2016

[CEP16] Conference Européenne des Administration des Postes et des Télécommunications, www.cept.org, 2016

[CEP16a] http://www.erodocdb.dk/default.aspx, 2016

[CEN16] www.cenelec.eu, 2016

[CIT16] Comisión Interamericana de Telecomunicaciones, www.citel.oas.org, 2016

[Com16] computersight.com/communication-networks/internet-structure-and-topology

© Springer International Publishing Switzerland 2017

R. Beutler, *Evolution of Broadcast Content Distribution*,

DOI 10.1007/978-3-319-45973-8

[DMB16] en.wikipedia.org/wiki/Digital_multimedia_broadcasting, 2016

[DSL16] en.wikipedia.org/wiki/Digital_subscriber_line, 2016

[DVB15] DVB Commercial Module - Terrestrial, *A Long Term Vision for Terrestrial Broadcast*, Study Mission Report of CM-T, https://www.dvb.org/resources/public/whitepapers/cm1621r1_sb2333r1_long-term-vision-for-terrestrial-broadcast.pdf, 2015

[DVB16] www.dvb.org, 2016

[DVB16a] www.dvb.org/news/worldwide, 2016

[DVH16] en.wikipedia.org/wiki/DVB-H, 2016

[EBU14] European Broadcasting Union, *Availability of PSB Programmes on TV Distribution Platforms*, EBU Fact Sheet, tech.ebu.ch/docs/factsheets/ebu_tech_fs_tv_distribution_platforms.pdf, 2014

[EBU14a] European Broadcasting Union, *Delivery of Broadcast Content over LTE Networks*, EBU Technical Report 027, tech.ebu.ch/docs/techreports/tr027.pdf, 2014

[EBU14b] European Broadcasting Union, *Assessment of Available Options for the Distribution of Broadcast Services*, tech.ebu.ch/docs/techreports/tr026.pdf, 2014

[EBU15] European Broadcasting Union, *Net Neutrality*, EBU Policy Sheet, www3.ebu.ch/publications, 2015

[EBU15a] European Broadcasting Union, *Audience Trends Television 2015*, www3.ebu.ch/publications, 2015

[EBU15b] European Broadcasting Union, *Simulation Parameters for Theoretical LTE eMBMS Network Studies*, EBU Technical Report 034, tech.ebu.ch/publications/tr034, 2015

[ECC14] Electronic Communication Committee (ECC), *Long Term Vision for the UHF Broadcasting Band*, ECC Report 224, www.erodocdb.dk/Docs/doc98/official/pdf/ECCREP224.PDF, 2014

[ECO16] European Communications Office, www.cept.org/eco, 2016

[EFI16] ECO Frequency Information System, http://www.efis.dk, 2016

[Eri15] Ericsson, *5G Systems*, White Paper, www.ericsson.com/res/docs/whitepapers/what-is-a-5g-system.pdf, 2015

[Eri16] Ericsson, *5G Radio Access*, White Paper, www.ericsson.com/res/docs/whitepapers/wp-5g.pdf, 2016

[ETS09] European Telecommunications Standards Institute, *Digital Video Broadcasting (DVB); Framing structure, channel coding and modulation for digital terrestrial television*, ETSI EN 300 744 V1.6.1, 2009

[ETS15] European Telecommunications Standards Institute, *Digital Video Broadcasting (DVB); Frame structure channel coding and modulation for a second generation digital terrestrial television broadcasting system (DVB-T2)*, ETSI EN 302 755 V1.4.1, 2015

[ETS16] European Telecommunications Standards Institute, http://www.etsi.org, 2016

[EuA15] www.euractiv.com/sections/digital/deutsche-telekom-chief-causes-uproar-over-net-neutrality-319028, 2015

[EuC14] European Commission, *Landmark agreement between the European Commission and South Korea on 5G mobile technology*, europa.eu/rapid/press-release_IP-14-680_en.htm, 2014

[EuC14a] European Commission, *Report on the results of the work of the High Level Group on the future use of the UHF band*, ec.europa.eu/digital-single-market/en/news/report-results-work-high-level-group-future-use-uhf-band, 2014

[EuC14b] European Commission, *Challenges and opportunities of broadcast-broadband convergence and its impact on spectrum and network use*, ec.europa.eu/digital-single-market/en/news/challenges-and-opportunities-broadcast-broadband-convergence-and-its-impact-spectrum-and-0, 2014

[EuC15] European Commission, *Public consultation on Directive 2010/13/EU on Audiovisual Media Services (AVMSD) - A media framework for the 21st century*, ec.europa.eu/digital-agenda/en/news/public-consul tation-directive-201013eu-audiovisual-media-services-avmsd-media-framework-21st, 2015

[EuC15a] European Commission, *Fact Sheet: Roaming charges and open Internet: questions and answers*, europa.eu/rapid/press-release_MEMO-15-5275_en.htm, 2015

[EuC15b] European Commission, *Towards 5G*, ec.europa.eu/digital-single-market/en/towards-5g, 2015

[EuC16] European Commission, *EU's spectrum policy framework*, ec.europa.eu/digital-agenda/node/118, 2016

[EuC16a] European Commission, *Telecoms Rules*, ec.europa.eu/ digital-agenda/en/telecoms-rules, 2016

[EuC16b] European Commission, *Digital Single Market - Bringing down barriers to unlock online opportunities*, ec.europa.eu/priorities/digital-single-market/, 2016

[EuP00] European Parliament and Council, *Directive 2000/31/EC of the European Parliament and the Council on certain legal aspects of information society services, in particular electronic commerce, in the Internal Market (Directive on electronic commerce)*, eur-lex.europa.eu/legal-content/EN/TXT/PDF/?uri=CELEX:32000L0031&from =EN, 2000

[EuP02] European Parliament and Council, *Directive 2002/21/EC of the European Parliament and the Council on a common regulatory framework for electronic communications networks and services (Framework Directive)*, ec.europa.eu/digital-agenda/sites/digital-agenda/files/140framework_5.pdf, 2002

[EuP02a] European Parliament and Council, *Directive 2002/19/EC of the European Parliament and the Council on access to, and interconnection of, electronic communications networks and associated facilities (Access Directive)*, ec.europa.eu/digital-agenda/sites/digital-agenda/files/140access_1.pdf, 2002

[EuP02b] European Parliament and Council, *Directive 2002/20/EC of the European Parliament and the Council on the authorisation of electronic communications networks and services (Authorisation Directive)*, ec.europa.eu/digital-agenda/sites/digital-agenda/files/140authorisation_2.pdf, 2002

[EuP02c] European Parliament and Council, *Directive 2002/22/EC of the European Parliament and the Council on universal service and users' rights relating to electronic communications networks and services (Universal Service Directive)*, ec.europa.eu/digital-agenda/sites/digital-agenda/files/Directive 2002 22 EC_0.pdf, 2002

[EuP02d] European Parliament and Council, *Directive 2002/58/EC of the European Parliament and the Council concerning the processing of personal data and the protection of privacy in the electronic communications sector (Directive on privacy and electronic communications)*, ec.europa.eu/digital-agenda/sites/digital-agenda/files/24eprivacy_2.pdf, 2002

[EuP10] European Parliament and Council, *Directive 2010/13/EC of the European Parliament and the Council on the coordination of certain provisions laid down by law, regulation or administrative action in Member States concerning the provision of audiovisual media services (Audiovisual Media Services Directive)*, eur-lex.europa.eu/legal-content/EN/TXT/PDF/?uri=CELEX:32010L0013&from =EN, 2010

[ExC16] www.explainingcomputers.com/internet.html, 2016

[Fac16] info.internet.org, 2016

[FCC15] Federal Communications Commission, *FCC Adopts Strong, Sustainable Rules to Protect the Open Internet*, www.fcc.gov/document/fcc-adopts-strong-sustainable-rules-protect-open-internet, 2015

[FCC16] Federal Communications Commission, www.fcc.gov, 2016

[Fib16] en.wikipedia.org/wiki/Fiber_to_the_x, 2016

[FLO16] en.wikipedia.org/wiki/MediaFLO, 2016

[FOB16] www.nercdtv.org/fobtv2012/index.html, 2016

[Foo16] commons.wikimedia.org/wiki/File:Satellite_Footprint.png

[GSM14] GSMA Intelligence, *Understanding 5G: Perspectives on future technological advancements in mobile*, gsmaintelligence.com/research/2014/12/understanding-5g/451, 2014

[Goo14] googlefiberblog.blogspot.de/2012/04/construction-update.html, 2014

[Goo16] fiber.google.com, 2016

[Goo16a] www.google.com/loon, 2016

[Hbb16] www.hbbtv.org, 2016

[HDT16] en.wikipedia.org/wiki/High-definition_television, 2016

[Hua13] Huawei, *5G: A Technology Vision*, White Paper, www.huawei.com/5gwhitepaper/, 2013

[Hul16] Hulu, www.hulu.com, 2016

[IEE16] en.wikipedia.org/wiki/IEEE_802.11

[IHS16] IHS Technology, technology.ihs.com, 2016

[InP16] en.wikipedia.org/wiki/Internet_Protocol

[iPl16] British Broadcasting Corporation, *BBC iPlayer*, www.bbc.co.uk/iplayer, 2016

[IPS16] en.wikipedia.org/wiki/Internet_protocol_suite

[IRT16] www.irt.de/en/research/digital-networks/imb5.html, 2016

[ITU04] International Telecommunications Union, *Regional Radiocommunication Conference (RRC-04)*, www.itu.int/net/ITU-R/index.asp?category=conferences&rlink=rrc-04&lang=en, Geneva, Switzerland, 2004

[ITU06] International Telecommunications Union, *Regional Radiocommunication Conference (RRC-06)*, www.itu.int/net/ITU-R/index.asp?category=conferences&rlink=rrc-06&lang=en, Geneva, Switzerland, 2006

[ITU06a] International Telecommunications Union, *Final Acts of the Regional Radiocommunication Conference for planning of the digital terrestrial broadcasting service in parts of Regions 1 and 3, in the frequency bands 174–230 MHz and 470–862 MHz (RRC-06)*, www.itu.int/pub/R-ACT-RRC.14-2006,Geneva,Switzerland, 2006

[ITU15] International Telecommunications Union, *Radio Regulations*, www.itu.int/pub/R-REG-RR/en, 2015

[ITU15a] International Telecommunications Union, *IMT Vision – Framework and overall objectives of the future development of IMT for 2020 and beyond*, Recommendation M.2083, www.itu.int/rec/R-REC-M.2083-0-201509-I/en, 2015

[ITU16] International Telecommunications Union, www.itu.int, 2016

[JTC16] portal.etsi.org/tb.aspx?tbid=92&SubTB=92, 2016

[Ken14] R. Kenny, R. Foster, T. Sute, *The value of Digital Terrestrial Television in an era of increasing demand for spectrum*, www.digitaluk.co.uk, 2014

[Mar15] MarketingCharts, *Are Young People Watching Less TV? (Updated – Q2 2015 Data)*, www.marketingcharts.com/television/are-young-people-watching-less-tv-24817, 2015

[Max16] Maxdome, www.maxdome.de, 2016

[Nas16] National Aeronautics and Space Administration (NASA), *Global Change Master Directory*, gcmd.nasa.gov/add/ancillaryguide/platforms/orbit.html

[NeF16] Netflix, www.netflix.com, 2016

[NeN16] en.wikipedia.org/wiki/Net_neutrality, 2016

[Ofc15] Office of Communications, *The Communications Market Report*, stakeholders.ofcom.org.uk/binaries/research/ cmr/ cmr15/CMR_UK_2015.pdf, 2015

[Ofc16] Office of Communications, www.ofcom.org.uk, 2016

[OSI16] en.wikipedia.org/wiki/OSI_model, 2016

[Rat16] D. Ratkaj, *Modelling the distribution value chain*, EBU Technical Assembly, https://tech.ebu.ch/docs/events/TA16/presentations/15_Modellingdistributionvaluechain_TA2016_rev1.pdf?ticket=ST-88519-SD50ybVXPRshwuH91R95-cas, 2016

[RCC16] Regional Commonwealth in the Field of Communication, http://www.en.rcc.org.ru, 2016

[Ric15] Richter, M.S., *Next Generation DTV: ATSC 3.0, International Symposium on the Digital Switchover*, www.itu.int/en/ITU-R/GE06-Symposium-2015/Pages/default.aspx, Geneva, Switzerland, 2015

[RSC16] Radio Spectrum Committee, ec.europa.eu/digital-agenda/en/radio-spectrum-committee-rsc, 2016

[RSC16a] Radio Spectrum Committee, circabc.europa.eu/ faces/jsp/extension/wai/navigation/container.jsp, 2016

[RSD02] European Parliament, *Decision No 676/2002/EC of the European Parliament and of the Council of 7 March 2002 on a regulatory framework for radio spectrum policy in the European Community (Radio Spectrum Decision)*, eur-lex.europa.eu/legal-content/EN/TXT/?uri=CELEX%3A32002D0676, 2002

[RSP16] Radio Spectrum Policy Group, ec.europa.eu/digital-agenda/en/radio-spectrum-policy-group-rspg, 2016

[Rua15] Ruane, K.A., *Net Neutrality: Selected Legal Issues Raised by the FCC's 2015 Open Internet Order*, www.fas.org/sgp/crs/misc/R43971.pdf, 2015

[San07] en.wikipedia.org/wiki/San_Francisco_Municipal_Wire less, 2007

[Sat16] en.wikipedia.org/wiki/Satellite, 2015

[SDT16] en.wikipedia.org/wiki/Standard-definition_television, 2016

[Sia16] www.siano-ms.com, 2016

[SpB16] en.wikipedia.org/wiki/Spot_beam, 2016

[SWR11] de.wikipedia.org/wiki/Alpha_0.7_-_Der_Feind_in_dir, 2011

[SWR16] Südwestrundfunk, www.swr.de, 2016

[TDF15] www.tdf-group.com/sites/default/files/PressReleaseTDF_FirsttrialLTEA_130415.pdf, 2015

[TNO14] TNO, *Regulation in the converged Media-Internet-Telecom value web*, www.tno.nl/en/focus-area/industry/networked-information/information-creation-from-data-to-information/regulation-in-the-media-internet-telecom-value-web/, 2014

[TrS16] en.wikipedia.org/wiki/MPEG_transport_stream, 2015

[UHD16] en.wikipedia.org/wiki/Ultra-high-definition_television, 2016

[UPU16] Universal Post Union, www.upu.int, 2016

[VDS16] en.wikipedia.org/wiki/Very-high-bit-rate_digital_subscriber_ line, 2016

[Vec16] en.wikipedia.org/wiki/Very-high-bit-rate_digital_subscriber_ line_2#Vectoring, 2016

[Ver14] www.verizonwireless.com/news/article/2014/01/lte-multicast-verizon-power-house.html, 2014

[WIP15] World Intellectual Property Organization, *Current Market and Technology Trends in the Broadcasting Sector*, www.wipo.int/edocs/mdocs/copyright/en/sccr_30/sccr_30_5.pdf, 2015

[WuT03] Wu, T., *Network Neutrality, Broadband Discrimination*, Journal of Telecommunications and High Technology Law, Vol. 2, p. 141, 2003

[ZDF16] ZDF, *ZDFmediathek*, www.zdf.de/ZDFmediathek, 2016

[Zen15] ZenithOptimedia, *Media Consumption Forecasts 2015*, zenithmedia.se/wp-content/uploads/2015/05/Media Consumption Forecasts 2015.pdf, 2015

[3GP16] www.3gpp.org, 2016

The list of references given here has to be considered as a subjective subset of those documents that might be relevant. It is by no means exhaustive. Rather, it reflects the author's knowledge and usage of literature. It might give first indications and hints for further reading. Additional references can be found with the help of any Internet search engine without problems.

Some of the references given point to documents that are not officially available. However, in most cases these documents are not classified and can be obtained by addressing the organisations or the authors directly.

Index

© Springer International Publishing Switzerland 2017
R. Beutler, *Evolution of Broadcast Content Distribution*,
DOI 10.1007/978-3-319-45973-8

Printed in the United States
By Bookmasters